SEAWEEDS

OF CAPE COD AND THE ISLANDS

JOHN M. KINGSBURY
Cornell University, and

PHILIP SZE
Department of Biology, Georgetown University

Illustrated by

EDWARD and MARCIA NORMAN

Bullbrier Press

JERSEY SHORE, PENNSYLVANIA

SECOND EDITION, REVISED
ISBN: 0-9612610-7-2.

Printed on acid-free, permanent paper and bound in the United States of America by Cayuga Press, Ithaca, NY 14850.

This book is available from Bullbrier Press, RR1, Box 332, Jersey Shore, PA 17740; FAX 717 769 7345.

Books by John M. Kingsbury
POISONOUS PLANTS OF THE UNITED STATES AND CANADA
DEADLY HARVEST - A GUIDE TO COMMON POISONOUS PLANTS
THE ROCKY SHORE
200 CONSPICUOUS, UNUSUAL, OR ECONOMICALLY IMPORTANT TROPICAL PLANTS
HERE'S HOW WE'LL DO IT; CONSTRUCTION OF THE SHOALS MARINE LABORATORY

Books by Philip Sze
A BIOLOGY OF THE ALGAE

Illustrated by Edward and Marcia Norman
A BEACHCOMBER'S BOTANY
THE SANDY SHORE
THE ROCKY SHORE
THE ATLANTIC SHORE
THE WINTER BEACH
BIRDS OF CAPE COD, MASSACHUSETTS

Library of Congress cataloging-in-publication:

Library of Congress Cataloging-in-Publication Data

Kingsbury, John Merriam, 1928-
Seaweeds of Cape Cod and the islands / John M. Kingsbury and Philip Sze ; illustrated by Edward and Marcia Norman. -- 2nd ed., rev.
p. cm.
Includes bibliographical references and index.
ISBN 0-9612610-7-2 (alk. paper)
1. Marine algae--Massachusetts--Cape Cod--Identification. 2. Marine algae--Massachusetts--Martha's Vineyard--Identification. 3. Marine algae--Massachusetts--Nantucket Island--Identification. 4. Botanical illustration. I. Sze, Philip, 1945- . II. Title.
QK571.5.M4K55 1997
579.8'17745--dc21 97-3931[illegible]
CIP

Dedicated to the lasting memory of

WILLIAM RANDOLPH TAYLOR 1895 - 1990

who for three quarters of a century as scholar, scientist, and teacher, examined the marine algae of the North American East Coast, who, for almost as long made Cape Cod his summer home, and whose studies made possible a book such as this.

PREFACE

This book is about the seaweeds of Cape Cod, Marthas Vineyard, and Nantucket. It is written for the serious inquirer. One of its main purposes is to help the inexperienced collector identify seaweeds, with equipment no more complex than a hand lens. Technical language is used only in the instances where common English is not precise enough.

Because Cape Cod juts into the Atlantic at a latitude where cold and warm waters meet, its seaweed flora represents to a marked degree the intermixing of northern and southern species. This book should be useful, therefore, well beyond the geographic limits of its title.

With the exception of a very few rare forms, all specimens illustrated here have been collected in Cape Cod waters, mostly by the illustrators themselves. Identifications have been checked by examination of reproductive or cellular detail in whole mounts or in sections under the microscope as necessary to insure accuracy. Its originators hope this book will also, therefore, fill a helpful role among the resources of professional marine biologists.

Interspersed among the discussions of individual species on the following pages are brief essays on seaweeds as organisms, their habitats, ecology, relationships and human significance. These essays, taken as whole, are intended to give the reader a general knowledge of this major group of organisms, and of their adaptations for living at the shifting interface between land, air and ocean.

J. M. K. and P. S.

CONTENTS

MN

INTRODUCTION

Over one hundred years have elapsed since American botanists began taking a close look at our East Coast seaweeds. Almost from the time of these first studies, Cape Cod was recognized as particularly interesting in that it seemed to separate a northern flora of arctic affinities from a more temperate flora to the south. The northern assemblage of seaweeds extends down the East Coast until it meets the sand barrier of the north side of Cape Cod. A flora characterized by southern forms can be traced from the south side of Cape Cod to New Jersey where increasingly rock-free sandy shores limit the number of seaweeds which can be found.

Anyone who swims on both sides of the Cape will realize that the water temperatures are quite different. The Gulf Stream bends eastward offshore not too far south of the Cape (some eighty miles off Nantucket) and this influences southern shore temperatures. Even more important in determining water temperatures are the comparatively shallow waters of Nantucket and Vineyard Sounds and Buzzards Bay, which warm readily in summer. The northern shore of the Cape, on the other hand, is bounded even in summer by the same cold water that bathes the shores of the Gulf of Maine. In winter, water on either side of the Cape cools to the same, nearly freezing, temperature. Thus, the water along the Cape's southern shore and around the Islands undergoes greater seasonal change in temperature and becomes much warmer in summer than does the water of Cape Cod Bay. These differences are reflected in the populations of seaweeds at each place and account in large part for the unusual richness of forms to be found by observers on Cape Cod and the Islands.

Seaweeds are marine algae

Natural waters, whether fresh or marine, contain many types of photosynthetic organisms. These are to be found floating, immersed, or attached to the bottom, and range in size from microscopic to massive. Principal among photosynthesic inhabitants of natural waters are the algae. Freshwater algae include pond scums, water blooms, floating blanket algae, freshwater plankton, and the like. The marine environment harbors floating marine plankton, attached seaweeds, and *Sargassum* of the Sargasso Sea. Traditionally, photosynthetic

Zostera marina "Eel grass" X 0.25

organisms have been called plants. Evolving scientific concepts of the diversity and complexity of the organisms grouped under that term bring into question its use for the seaweeds or algae generally. Seaweeds is a poor term, also, for the marine algae because they are not weeds in the same sense as garden weeds. When we use the terms "seaweeds" or "plants" for marine algae in the following text, consider them used loosely.

Algae and plants are the principal photosynthetic organisms on our planet. In contrast to true plants, algae lack such specialized structures as flowers, seeds, and vascular tissue, which evolved for life on land. Rather the algae are a very diverse assortment of organisms, spanning the range of complexity from bacteria to plants and defying a simple definition. In fact, many scientists believe that different ancestral cells may have independently acquired photosynthetic systems and then evolved into today's algae. If that is so, the algae are an artificial grouping based only on the lack of any distinct charactertistics that would fit them into other groups.

Eelgrass is not a marine alga

Very few flowering plants live among the true seaweeds. The most common is eelgrass *(Zostera marina)*. This plant is genuinely marine, growing on muddy bottoms in protected waters from low tide to a depth of twenty feet or more, and forming dense populations many acres in extent. It can be found from Greenland to South Carolina. Populations of eelgrass are important in providing food and protective cover to a complex association of animals, and the eelgrass blades serve for the attachment of many small seaweeds. At the latitude of Cape Cod, practically no seaweeds can attach to or grow from a sandy or muddy bottom. Thus without this plant, areas now covered by eelgrass would be like a barren desert. In 1931, a serious disease struck the eelgrass beds and in succeeding years wrought such devastation that some thought the eelgrass, like the American chestnut, would be essentially wiped out. Fortunately, the eelgrass developed resistance and the beds have slowly recovered. Great masses of eelgrass sometimes wash ashore, especially after onshore storms, and form mattress-like windrows as the tide retreats. These are bothersome on swimming beaches, especially when they decay rapidly (though this is not usual). Some individuals, not realizing the importance of this plant in the successful existence of fish and birds, would like to destroy the eelgrass beds, and extensive raking by clammers has the same effect.

To recognize eelgrass, look for thin leaves with parallel edges and three principal veins running lengthwise. The leaves, often nearly a yard long, are green as they grow, turn black when they die, and then may bleach white as they lie in tangles above high tide line. In nearly-transparent living leaves you may observe small cross-lines or checks when you hold them to the light. The flowers are small, inconspicuous, short-lived, and are produced only after the plant is several years old. Inflorescences, as shown enlarged in the illustration, ripen in early spring and summer and are not likely to be found by the late summer collector. The seeds are used as food by marine birds, especially the brant, Canada geese and wild ducks, and by some fish.

Lomentaria clavellosa on *Chondrus* X 1.0

Where to look for seaweeds.

Almost all living seaweeds need something solid for attachment; commonly rock, wood, shells, or other seaweeds. The best place to look for them, then, is on a rocky shore. Cape Cod is essentially a large ridge of sand and gravel scraped from other parts of New England and deposited at the edge of the last glacier to cover North America when it retreated 10,000 years ago. Marthas Vineyard and Nantucket were deposited at the edge of a somewhat earlier glacier. What we see today results from the combined effects of weather, wind, and water on what was left by the glaciers. The glaciers carried some surprisingly large rocks from as far away as the granite coast of Maine. Boulders twenty or thirty feet in circumference and many tons in weight are not uncommon. The larger the rock, the less distance has it moved from the place where the glacier left it.

The pile of coarse gravel and boulders concentrated at the temporarily stationary edge of a glacier is called a terminal moraine. Terminal moraines follow the high lands of the northwest and northeast sides of Marthas Vineyard, and the north and northwest sides of Nantucket. On Cape Cod a terminal moraine follows the east side of Buzzards Bay from the Elizabeth Islands to the Cape Cod Canal, thence more or less inland along the north side of the Cape to Chatham, the "elbow", then northerly to Provincetown. The best place to look for seaweeds is where the terminal moraine meets the ocean, for here the larger rocks have been concentrated as the sand has been washed away. Such places occur all along the east shore of Buzzards Bay; on the Cape Cod Bay shore from Barnstable Harbor to the turn north at the "elbow"; on the Islands as suggested above, and in other isolated locations. Nobska Point, just below the lighthouse to the east of Woods Hole is particularly productive because here the rocks are washed by a rapid current of water passing back and forth from Buzzards Bay to Nantucket Sound through Woods Hole (a nautical term, not the town) as the tides change.

Jetties, breakwaters, groins, and rock wharves are also good collecting spots and occur throughout Cape Cod waters. Care must be taken on some of these to avoid tangles and fish hooks. Barnacle shells have razor sharp edges and can cause painful cuts, slow to heal. If opportunity presents, it is particularly instructive to follow the sequence of colonization of a new groin by seaweeds and invertebrates. Several years may be required before the populations reach a steady-state or climax.

Algae have not been as well studied by scientists as have land plants. Keenly-observing naturalists can still make original discoveries of importance. For example, the Normans found *Lomentaria clavellosa* growing on the outer Cape. Previously this European seaweed (shown to the left) had been reported only once each from the vicinity of Boston and Gay Head. Biologists know little about which marine animals eat what seaweeds, the particular conditions that lead to the appearance of specific organisms at specific times and places, how complex life cycles are completed, or what happens to the summer seaweeds in winter. Careful observers who are in the right places at the right times can help in matters like these.

Tides and currents

Cape Cod waters are among the most interesting in the world in relation to tides and tidal currents. The average vertical distance between high and low tides is nine to ten feet along the north shore of the Cape at Sandwich, Barnstable, Wellfleet, and Provincetown, but only two to three feet at locations throughout Nantucket and Vineyard Sounds. At Gay Head, Cuttyhunk, New Bedford, and the lower part of Buzzards Bay generally, it is between three and four feet, while it increases somewhat along the shores of Buzzards Bay north to Wareham. Furthermore, low and high tides come at a different time in Buzzards Bay than at other locations. Tides at most places in the Bay are approximately 3¼ to 3½ hours ahead of the rest of the Cape and the Islands. Low tide at West Falmouth Harbor, for example, is about 3½ hours earlier than low tide at Falmouth Heights.

These peculiar tidal characteristics have at least two consequences of importance to seaweed collectors. First, to do serious collecting on the northern shore of the Cape, or the outer shore north of Chatham, you need to know when low tide occurs. With an average vertical difference of some ten feet, and a difference during spring tides of almost twelve feet between high and low tides, most seaweeds are not well exposed except near low tide. On the other hand, the minimal tides along the south shore of the Cape and at the Islands allow you to do some collecting almost anytime. Furthermore, if you are in the Falmouth area, you have two choices for low tide, three hours apart, and can even do both on the same day. Where else in the world can you do that?

Secondly, these tidal peculiarities produce very strong tidal currents in some places. Strong currents not only are dangerous for the snorkeling or SCUBA-diving collector, but also are of considerable significance in the life and distribution of a seaweed. Consider what happens when you have two connected masses of water in which low and high tides differ in height and in time. Such a place occurs between Buzzards Bay and Vineyard Sound. The tidal difference between Nobska Point on the Sound, and the northwest side of Penzance Point, on the Bay, within walking distance of each other on either side of the town of Woods Hole, is more than two hours. High tide at Penzance may be more than two feet higher than high tide at Nobska. This means that water in Buzzards Bay on one side of Woods Hole is rarely at the same level as water of Vineyard Sound on the other side. Therefore strong currents, flowing up to eight miles per hour, wash through Woods Hole first one way and then the other. This cycle is repeated each day with the two high and low tides.

The seaweeds at Nobska Point benefit from the renewal of nutrients, removal of wastes, and mixing of water as it washes past them. In winter they suffer from the effects of ice scraping back and forth past the rocks. The same is true at Robinsons Hole and Quicks Hole which also join Buzzards Bay and Vineyard Sound through the chain of the Elizabeth Islands. Seaweeds on the rocks lining the Cape Cod Canal are exposed to currents nearly as great, occasioned by similar differences in time and height of tides in Buzzards Bay and Cape Cod Bay which it connects.

The differences between spring and neap tides are important in determining the distribution of seaweeds where the tidal rise and fall is considerable. The general characteristics of tides are determined primarily by the moon and sun through gravitational effects. The specific characteristics of the tidal rise and fall on a given shore are often influenced nearly as much, or even more, by the configuration of the shore, the inherent oscillation frequencies of the bays and the open ocean bounding them, the slowing effect of shallow bottoms and many similar considerations. Modifying influences of this sort are responsible for the three hour difference in tidal timing between Buzzards Bay and Nantucket Sound, described above.

Looking only at the general effects of moon and sun, certain characteristics of tides along the north shore of the Cape and elsewhere can be explained. High and low tides occur twice a day (semidiurnal tides) along Cape Cod's shores and in most parts of the world. The moon's gravitational pull causes the earth's water to bulge out on the side toward it, and at the same time almost equally, on the side of the earth away from it where its effects are therefore minimal. The earth rotates once each 24 hours beneath these two standing waves. But the moon simultaneously advances in orbit. The moon rotates around the earth once every 28 days. Therefore a given tide occurs 1/28th of a day (or about 50 minutes) later each day. In this orbital passage, the moon lies between the earth and the sun once every four weeks, and the earth lies between the moon and the sun once every four weeks, two weeks later. On these occasions, i.e., every two weeks, the moon and sun pull together on the earth's waters, either from the same side or from opposite sides, making the high tides extra high and the low tides extra low. these are called spring tides. A week later, the moon and sun are at right angles to each other in respect to the earth and their gravitation does not pull together. High tide is not very high, and low tide is not very low. These are called neap tides. To put this in specifics, the vertical distance between a spring high and low at the Cape Cod Canal jetty at Sandwich is about eleven feet. The difference between a high and low neap tide is about eight feet. There are two low tides each day. Tomorrow's tide is nearly an hour later than today's. The best collecting is during the lows of a spring series every two weeks.

Collecting equipment

Some collectors will be fortunate enough to possess diving equipment that enables them to examine marine algae which live only in areas below the low tide line. On the other hand, seaweeds from underwater rocks break free from time to time and wash ashore. Thus, most seaweeds that can be collected by diving can also be found, sooner or later, in good condition floating to shore at the edge of the water on a rising tide. After collection, seaweeds should be kept moist until pressed. It is more important to keep them cool and shaded than to keep them covered by water. Specimens may be collected into a pail with a small amount of water, or into a plastic bag. Small plants, otherwise easily lost, should be segregated into individual containers such as small plastic bags. Glass bottles or vials are dangerous on slippery rocks. A putty knife is useful in detaching some of the larger seaweeds from the rocks where they grow. Finally, a hand lens of about 10x magnification is necessary for careful collecting. You should buy either a very cheap one that can be discarded when it has rusted too much, or an expensive one which will neither rust nor leak seawater between the lenses. Field gear, thus, need not be extensive, and lack of any item should not deter a person from collecting. Serviceable plastic containers (unfortunately) can be picked up on almost any shore; aluminum beer cans are ubiquitous. Quahaug shells make reasonable substitutes for a putty knife.

Remember when collecting specimens to practice environmental responsibility by taking only what you need, always leaving enough algae to regrow and maintain the population.

Preserving your specimens

Seaweeds are easily preserved by drying on paper; in fact this is the principal method by which they are preserved for study by professional botanists. Use heavy, white paper that will withstand being immersed in seawater. Under a thin layer of water in a flat pan, or by means of a large medicine dropper, float the specimen out on top of the paper. Arrange its parts so that they show to their best advantage. You may prune away some parts if the material is too dense. Delicate plants may be teased out with needles (remember steel rusts rapidly in seawater).

Next, remove the paper from the water (or the water from the paper) in some way that will not disturb the specimen. Let it drip-dry, then finish drying it completely. This must be done rapidly so the plant will not begin to decay, and in such a way that the specimen does not curl the paper as it shrinks on drying. Botanists use a plant press and blotters. A weighted pile of warm newspapers works nearly as well. Use a whole folded newspaper between each two specimens. The specimen should be separated from the newspaper above it by a piece of cloth, otherwise it will stick to the newspaper as well as to the mounting paper. Relatively lint-free, fine-weave, color-fast or white, flat cloth such as old percale sheets is preferred. Coarse-weave cloth will leave an impression of the weave on soft seaweeds.

Be sure your newspapers are bone-dry before you start. Change them at least twice a day until the specimen is completely dry to the touch, then gently peel back the protective cloth. Cloths may be used repeatedly as long as they do not become seriously discolored by mucilage secretions from drying specimens. You may hasten the drying process by exposing the pile of newspapers to the moderate heat of a radiator or to bright sunshine. Most seaweeds will stick firmly to the mounting paper when dry. The few that do not adhere can be affixed later with ordinary white paste.

When you leave the moist seashore for a dryer climate, your specimens will continue to dry. Store them so they cannot curl. In a heated house in the winter some will become very brittle and must be handled gently. The stiffer the mounting paper, the less breakage will occur.

A satisfying hobby is making seasonal greeting cards from some of the more delicate seaweeds. Greens and reds appropriate to the midwinter season can be found in plants such as *Callophyllis*, young *Palmaria, Grinnellia, Dasya, Enteromorpha*, and *Ulva*. These algae also stick well to paper, handle reasonably easily, and give a good range of interesting form with which to work in making creative displays on paper. Keep cards out of direct sunlight after they have been prepared. The greens in particular fade quickly in light.

Photographing seaweeds

An alternative to collecting and pressing seaweed specimens is to photograph them either in their natural habitats or arranged in dishes or aquaria. An important precaution is to protect your camera from seawater—salt spray can quickly damage a camera. If you are going to photograph in the field, a sturdy plastic bag is an easy way to cover your camera, or alternatively use an "all-weather" or a one-use "disposable" camera. Good pictures of larger seaweeds can be taken with relatively simple point-and-shoot cameras. However, serious photographers with more sophisticated equipment may want to consider a macrolens for close-ups and a polarizing filter to reduce reflection from the water's surface. An "all-weather" camera can be fully submerged to its rated depth and will give interesting underwater views of seaweeds.

Another approach to photographing seaweeds is to float fresh specimens in a pan or to arrange them in an aquarium filled with seawater. Photographing outdoors in open shade will ensure adequate lighting and avoid spots of reflection from direct sunlight

Identifying seaweeds

Identifying seaweeds accurately is not easy. Authoritative reference texts provide technical descriptions and dichotomous keys for this purpose, but experience shows that the special terms and amount of detail in such works necessary for an inclusive and absolutely accurate treatment is often overwhelming for the beginner. This book relies instead on accurate illustrations with which specimens can be compared when collected. The illustrations are lifelike, and for most species, show a typical plant as it appears underwater. The gross structure and relationship of parts for the majority of seaweeds is best seen when the plant is in water. Taking it out of water causes the branches to collapse together into a structureless mass. Illustrations and descriptions are of plants as they appear to the naked eye or as magnified by hand lens.

The illustrations have been grouped in a logical sequence, first into major groups on the basis of color, and secondly within each major group, along lines of evolving complexity as presently understood. No keys have been included, but an outline which will reduce the number of illustrations to be searched in any given comparison will be found on page 211. For a long time, *Marine Algae of the Northeastern Coast of North America*, by William Randolph Taylor, was the authoritative manual for this area, but it has become out-of-date as taxonomists have reevaluated algal species and their relationships. Villalard-Bohnsack's *Illustrated Key to the Seaweeds of New England* is a recent authoritative key for identifying local seaweeds, but no taxonomic manual has become available to take the place of Taylor's book for the New England coast. However, there are several good reference manuals to adjacent areas, and these can provide information on specific species when more critical identifications are required. *Seaweed Flora of the Maritimes, 1. Rhodophyta - The Red Algae*, by Carolyn J. Bird and Jack L. McLachlan, has excellent photographs, and *Seaweeds of the Southeastern United States*, by Craig W. Schneider and Richard B. Searles, is useful for local species with warm water affinities.

The Green, Brown, and Red Seaweeds

Scientists have long divided marine algae into major groups on the basis of color. The first decision to be made in identifying a specimen is to which of these groups it belongs. This is not as easy as it sounds because healthy growing plants may contain pigments additional to the fundamental ones, and the resulting color may not be typical for their group. Furthermore you may have chosen a specimen that is past its prime and has begun to bleach or decay. Of the three main groups, green seaweeds are most consistent in color. Any seaweed which is grass-green almost surely belongs here. "Grass-green" means the green of a rapidly growing lawn in spring. Brown seaweeds, particularly the kelps and *Desmarestia viridis,* may take on a greenish cast as they decay, but this color is never without some hint of brown. Some greens, such as *Prasiola stipitata* or *Cladophora sericea* may be quite dark, almost grayish, *en masse*.

Brown seaweeds are less consistent in color. They may be bright yellow (shallow-water, sun-drenched), yellow-green-brown or olive (many species), black (some *Chordaria*), red-brown similar to winter oak leaves (old *Fucus*), or a true, medium brown (the majority). Red seaweeds are the most varied in color. Colors found in this group range from pink to wine red, purple, deep red, black-red, or black. Some, such as *Porphyra umbilicalis,* are a dull brown or olive. Red seaweeds tend to bleach or age through browns and yellows to white. Always check the bottom or most shaded part of a specimen before deciding to place it with the browns; a red seaweed usually shows its fundamental colors here. Some reds, especially *Chondrus,* iridesce with shimmering shades of purple when under water in sunlight. This disappears if the plant is removed from the water. Many delicate red seaweeds decay very rapidly, passing through a color phase which is almost fluorescent orange. If you see this color, the plant is too far gone to make an acceptable dried specimen.

Plant-like animals

Some invertebrates look like plants. The sea anemone takes its name from its petal-like tentacles, but is easily recognized as an animal by their muscular action or rapid retraction. Less easily recognized as members of the animal kingdom are those organisms previously called "bryozoan" (literally "moss-animals") but now known as "ectoprocts" (the common ones, at least). Some of these colonial

Sertularia sp. *X 0.25* *enlarged detail*

animals create horny crusts on seaweeds or other subjects. You will undoubtedly find one sooner or later on an Irish moss plant *(Chondrus).* Others form erect, leathery colonies that may reach an inch or two in length. *Sertularia* of this type, is illustrated. Occasionally, material of this sort is collected in various parts of the world, dyed green, and sold through chain stores as "live-forever fern" or under some similar designation. Of course it is neither alive nor is it a fern. *Sertularia* and its relatives are never distinctly green, deep brown, or red. Usually they are a bleached yellow-white, brown-white, or chalk white. The individual small animals live separately within tiny shells which are responsible for the color. You can just make out the individual shells and their openings with a handlens.

What to expect

Even using these illustrations carefully, you should not expect to be able to identity every seaweed you find. Seaweeds, like most living organisms, are variable. The specimen you collect may be quite different from the material we have illustrated. Our intent has been to illustrate typical material. In a few cases range of variation is shown, and in a few others we have explained how the illustrations may differ from what you may find. Your specimen may be very young or very old, or only part of a plant. It may be senescent and losing parts, especially the finer branchlets. It may have been washed in from very deep water, or it may have grown originally in a salt marsh or under some other unusual conditions. Or it may just be an odd individual, or so rare that it is not included here. In certain groups of seaweeds, microscopic examination is necessary to separate and identify species; for these, identification to species is beyond the scope of this book. The wise person will attempt, whenever possible, to look at several specimens of the same plant with an eye to recognizing what is typical before attempting to identify it. Recognition is more an art than a science. We hope with practice, and this book, you will be able to identify almost all of the seaweeds you find, at least to a good first approximation.

This is a book of illustrations, each of which would be important even if nothing were said about it. The essays accompanying each illustration give information about classification, derivation of genus name, major characteristics important in recognition, and typical habitat. In almost all cases each essay presents additional information. What has been included is not random, but follows a pattern designed to introduce the reader to a range of topics concerning seaweeds, including how and where they live, how they are built, how they reproduce, what roles they serve in the complex web of nature, their importance to humanity, and the particular characteristics of Cape Cod waters and shores as they relate to the seaweeds which populate them.

SEAWEEDS
OF CAPE COD AND THE ISLANDS

Calothrix sp. *zonation*

Calothrix crustacea

Derivation: "beautiful thread."
Range: entire East Coast.

Above the level submerged at high tide, rocks in exposed locations are often discolored by an obvious black, paint-like stain on top or on the faces toward the water. The black areas are irregular in outline and are determined in size by the size of the rock. Some may be several feet across or continuous, forming a band above the rockweeds and barnacles, as shown in this illustration. Each is composed of millions of microscopic threads (usually of *Calothrix crustacea)* of characteristics too small to be made out with a hand lens. Each thread, seen under a microscope, consists of a single file of cells. This is termed a filament. The aggregation is dry and powdery to the touch when the rock is dry; moderately slippery when wet. Dry material can be scraped off with a pocket knife and works out to a powder between the fingers.

Although there are many marine types of cyanobacteria, for the most part they cannot be recognized without resorting to a microscope, and hence are beyond the limits of this book. Formerly, cyanobacteria were called bluegreen algae, but after being examined with electron microscopes and found to lack an organized nucleus and certain other cellular components found in more advanced organisms (including algae), they are now classified as a group of bacteria. Their photosynthetic pigments are not contained in chloroplasts within the cells as in other algae, and their reproduction is by simple cell division and fragmentation.

Cyanobacteria grow on various substrates. Commonly they form a microscopic fuzz on the surface of other seaweeds. Some pioneer on the surface of marine muds, and start the succession that leads to eventual colonization by higher plants. Sometimes under favorable circumstances, mats of cyanobacteria start on the bottom of shallow, warm bays, and eventually float to the surface as thread-like skeins, up to a yard long, of purple or black-green color. Some years the bay between Bassetts Island and Scraggy Neck, for example, is filled with them for a brief period in late summer.

If you find a thick, dull green film resembling a waterlogged green window shade on sod or mud near high tide line, it is apt to be a species of *Vaucheria* (named after Jean Pierre Voucher, a Swiss botanist). This is the only representative that we will mention of another major group of algae, the yellow-greens (Division Chrysophyta, Class Tribophyceae), most of which are found in the freshwater environment.

Ulothrix flacca X 1.0

Ulothrix flacca

Ulotrichales

Derivation: "forest of hairs."
Range: South Carolina northward.

This seaweed is gregarious, creating populations that resemble green whiskers up to an inch or so long on rocks and pilings near the high tide level. Often it forms a distinct zone of growth just below the *Calothrix* level. We have collected it also growing on *Fucus* plants at the Sandwich jetty (Cape Cod Canal entrance). *Ulothrix* is present during the colder months of the year and normally disappears during summer.

Each *Ulothrix* plant in the population is a simple, unbranched filament or thread of cells attached end to end in single file, the bottom cell in turn attached to the substrate. Spread a small tuft out in a drop of water on your fingertip. Under a hand lens you should be able to determine that the filaments are very fine and unbranched. Individual cells are too small to see by hand lens. If your collection looks at all ribbon-like, you probably have *Blidingia*. If you can see individual cells, you probably have a member of the Cladophorales (p. 41). If it is red to black in color, try *Bangia* (p. 127).

Each cell of the filament contains a well-defined nucleus and a membrane-bounded chloroplast in which the photosynthetic pigments are located. The electron microscope shows that the pigments are not simply dissolved in this sack-like structure, but instead are components of delicate membrane layers. The principal characteristic unifying the green algae is the presence in the chloroplast of a characteristic assortment of photosynthetic pigments including chlorophylls and carotenoids of specific types. The result is a typical grass-green color if additional, non-photosynthetic pigments do not obscure or change it.

The use of color to separate groups of algae, dating from about 1800, was perhaps the first application of a biochemical characteristic at so fundamental a level in classification. Fortunately, it has stood the test of time. Additional characteristics of structure and reproduction as they have been worked out in succeeding years, have supported the original scheme. For example, swarmers (a general term for the microscopic motile reproductive cells of algae) of *Ulothrix* bear near one end two or four whip-like protoplasmic extensions called flagella (singular, flagellum). By flexing and beating rapidly, these serve to propel the swarmer through the water. The swarmers of Green Algae (Division Chlorophyta) can be characterized as possessing two flagella each, or a multiple of two, and having flagella that are always identical in structure. In other groups, the swarmers have flagella unequal in number or structure.

Monostroma oxyspermum X 0.50

Monostroma oxyspermum Ulvales

Derivation: "single-sheet"
Range: entire East Coast

Monostroma starts its growth as a tiny balloon whose wall is a sheet of cells one cell thick (whence the name). Soon it splits open and the sheet continues diffuse growth as a blade. The plant shown here is large and fully mature. It is an inhabitant of shallow water and protected exposures and may become particularly abundant in waters of less than full salinity where it may form extensive populations. Kingsbury has collected it from rocks in the outfall from a freshwater pond near Old Silver Beach (Falmouth) at the level where extreme high tide meets the outflowing fresh water only briefly twice a day. It is a plant of spring and summer, disappearing later, to reappear the following year.

Each recognizably separate type of plant (or animal) is considered a separate species. Species showing sufficient similarity or relationship are grouped into genera (singular, genus). Each species is therefore designated by two names, the generic first and the specific second. Naming of organisms is controlled by international agreement and regulation. Names and descriptions are expressed in Latin because as a "dead" language it is not subject to change in meaning with use. Any given combination of genus and species names can legitimately be used for only one organism. Most scientific names for seaweeds either are based on Latin or Greek roots that are somehow descriptive of the plant, or are given in honor of a particular person, often for a particular reason. The derivation of each genus name is given in this book as an aid to remembering them.

The generic name of *Monostroma* is about two centuries younger than the oldest houses on Cape Cod. In order to stabilize the science of putting names on plants, botanists have chosen a book by the Swedish botanist Linnaeus, written in 1753, as the starting point for all plant names. Linnaeus was one of the first persons to use a consistent system of names for plants and he preserved specimens (which can still be studied) of the plants he described and named. *Ulva* is a Linnaean name, dating therefore to 1753. As Linnaeus used it, *Ulva* included all the blade-like green seaweeds. *Monostroma* was recognized as different from *Ulva*, and named as a separate entity almost exactly a century later.

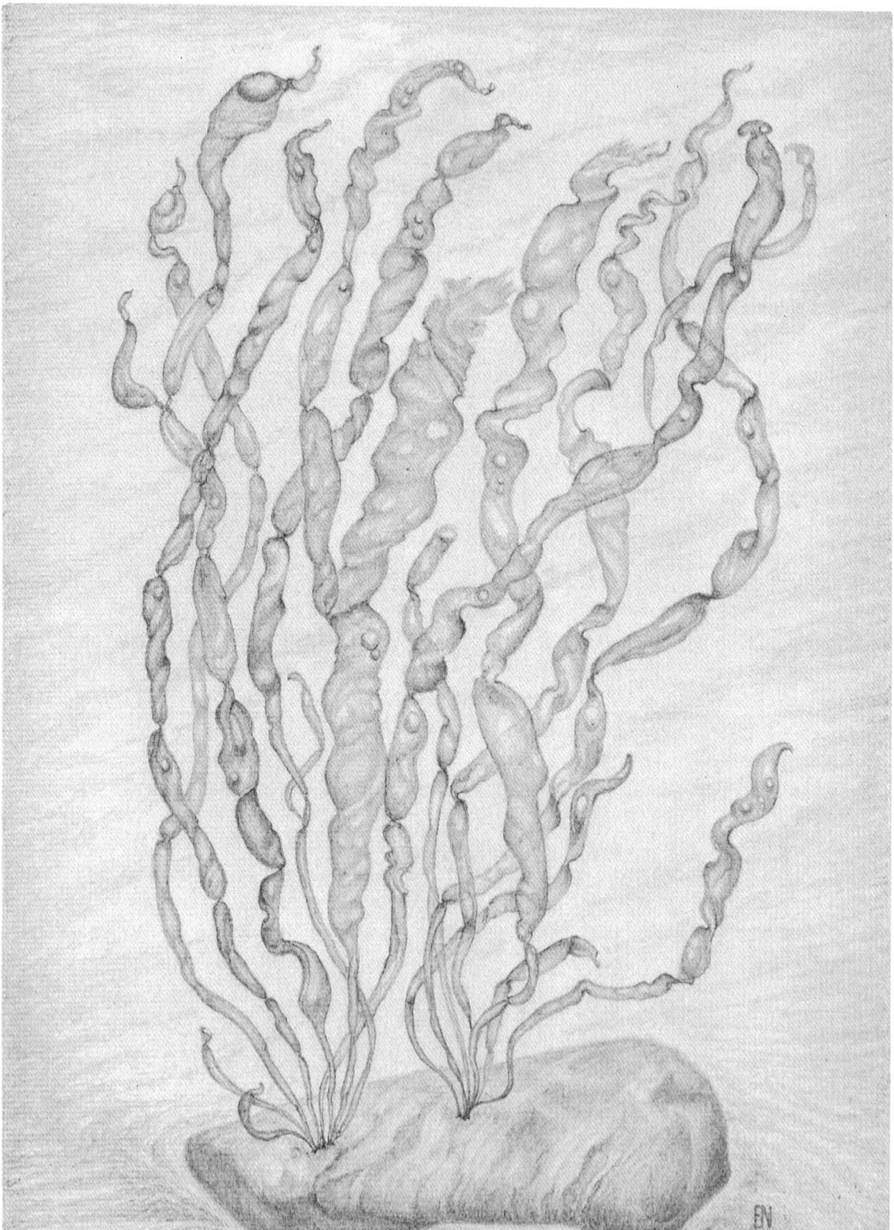

Enteromorpha intestinalis X 1.0

Enteromorpha intestinalis Ulvales

Derivation: "gut-shaped."
Range: entire East Coast.

This seaweed is perhaps as well named as any. Its generic name means "gut-shaped" and its species name means "intestine-like," as though to emphasize the point. Few seaweeds have attracted enough popular attention to have earned common names, and no attempt has been made to coin artificial common names for them in this book.

Enteromorpha intestinalis is widely distributed along our Atlantic Coast and on European shores. Typically it is found between tide lines in protected waters of harbors and coves. It attaches readily to stones, shells, wood, and other solid objects and usually forms conspicuous populations. An average, mature plant may be a foot or more long and half an inch or so wide. It is a mostly unbranched, bright green (or somewhat bleached) tube. Under certain conditions, especially under warm sunny conditions which promote photosynthesis, the hollow tube becomes irregularly puffed out or inflated with gas. Floating in tide pools or on the incoming tide, a group of these plants looks very much like a tangle of intestines from some small animal, except for color.

The clusters of plants shown in the illustration are entirely under water and contain some obvious gas bubbles. This causes each tube to extend strongly upward in the water. The walls of the tubes are only one cell thick and the upper ends appear frayed and worn after the cells have become reproductive.

Occasionally, *Enteromorpha intestinalis* will fill a tidepool near the high tide level. In these circumstances it is occupying space and preventing other seaweeds from growing in the pool. Studies by Dr. Jane Lubchenco on Cape Ann (Masssachusetts) showed that the composition of the seaweed communities of tidepools located high in the intertidal zone is regulated by the feeding activity of the herbivorous Common Periwinkle (*Littorina littorea*). The periwinkle has a definite preference for *E. intestinalis* and by feeding on it lessens its ability to monopolize the space in a pool. The result is that a more mixed assortment of seaweeds can occupy the pool, none of which has as strong a competitive advantage as *E. intestinalis.* As you wander along the shore, look at pools near the high tide level and see if they follow this pattern: pools in which *Enteromorpha* is abundant and periwinkles are absent, and pools in which periwinkles and other seaweeds are common and *Enteromorpha* is sparse.

Enteromorpha flexuosa X 1.0

Enteromorpha flexuosa Ulvales

Derivation: "gut-shaped."
Range: entire East Coast.

Various species of *Enteromorpha* share the same basic structure and reproduction. They differ in such things as habitat, branching, size and arrangement of cells, and chloroplasts. Many species have been recognized in this genus and they are not easily separated, even on technical grounds.

Plants of *Enteromorpha*, branching or not, are composed of a single sheet of cells in tube form. The cells are like those of *Ulothrix*, but instead of dividing in only one plane so that a filament results, here they divide in two planes and a sheet of cells results. Germlings of *Enteromorpha* go through a few-celled *Ulothrix* stage before they broaden out and become tubular. Growth is diffuse; any cell can divide, and division appears more or less random under the microscope. That it is not random is obvious from the fact that gross characteristics of the plant are definite. *Enteromorpha* is attached by means of the downward growth of microscopic colorless projections from the lowermost cells of the tube that adhere to the substrate.

When you pick up an *Enteromorpha* its cells may contain a single complement of chromosomes (haploid condition) or, like the cells of our own bodies, a double complement (diploid). The two types of plants alternate; each one, when reproducing, generates the other. In either case, reproduction is by means of swarmers—tiny microscopic swimming cells produced severally in each cell of the parent. All the contents of the parent cell is used up in the formation of the swarmers. They are released at maturity through a pore that develops in the cellulosic wall of the parent cell, leaving only dead cell walls behind. Each swarmer bears two or four hairlike whips or flagella that beat vigorously and propel the swarmer through the water. Swarmers from diploid plants (zoospores) swim for a while, then settle and attach to some solid surface, and grow into a haploid plant. A reduction division of the chromosomes, one of the two major cellular events in the life history of any sexually reproducing organism, has taken place in their formation. Swarmers from haploid plants (gametes) fuse in pairs, thus again doubling the number of chromosomes, then settle, attach, and grow into the diploid generation. Both generations are identical in structure. This is one of the basic types of life cycle among algae and is diagrammed on p. 33.

Enteromorpha flexuosa, shown here attached to a broken, eroded shell of the whelk or conch *(Busycon)*, grows between tides or below low tide in moderately exposed locations. The plants are soft and the tubes slender. Careful observation will show that the major tubes bear many smaller lateral branches.

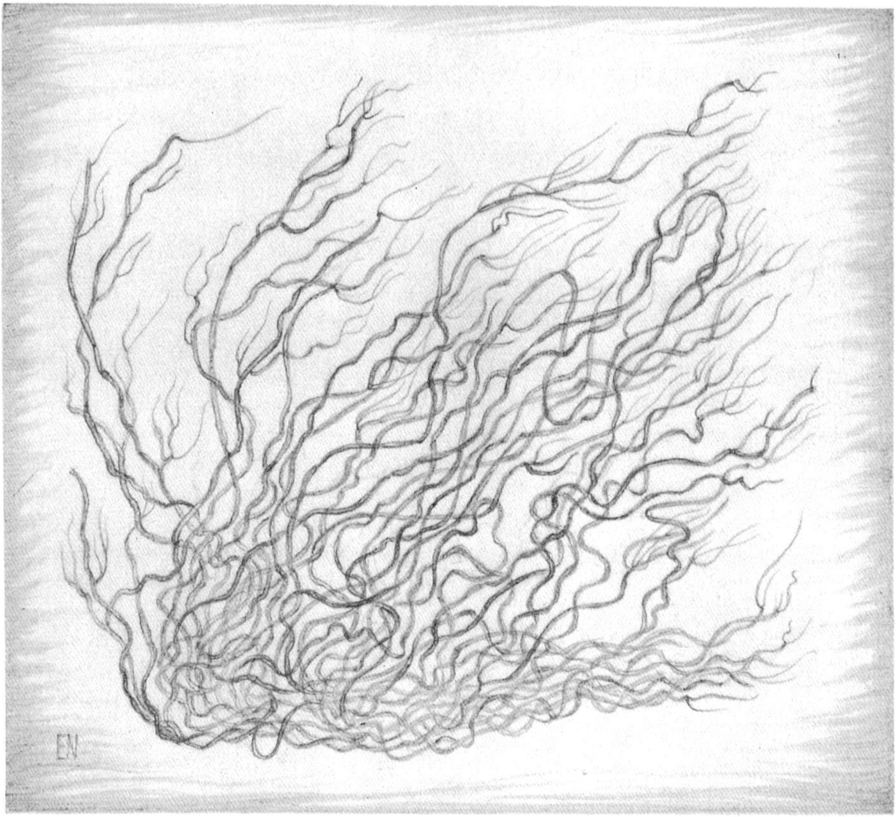

Enteromorpha clathrata X 1.0 *portion of entangled mass*

Enteromorpha clathrata *early growth* X 1.0

Enteromorpha clathrata

Ulvales

Derivation: "gut-shaped."
Range: entire East Coast.

Though its center of abundance is south of Cape Cod, *Enteromorpha clathrata* ranges from Brazil to the Arctic. The seaweed is composed of fine, branching, cylindrical tubes whose surfaces are sheets of cells one cell thick. It is found characteristically in shallow bays or salt marshes in summer where the water is warm and undisturbed by waves. Under these conditions it grows rapidly and indefinitely, forming extensive green or yellow-green mats that lie on the bottom or loosely entangled among the eelgrass leaves or other plants, or may float at the surface. Each individual plant starts out attached to some solid surface, but as the tangles become extensive, they usually break free from the original attachment. Unlike most brown and red seaweeds this does not prevent them from continuing growth. Under good growing conditions, the mats of tangled branches may become very extensive, covering areas yards or more in extent.

Comparison of the four species of *Enteromorpha* discussed here will give an idea of the type of similarity which serves to join species as members of a single genus. A closely related genus *Blidingia* (mentioned under *Ulothrix*) is a small version of *Enteromorpha* (usually less than an inch long) that has a distinctive button-shaped holdfast for attachment to the substrate. Comparison of *Enteromorpha* with its relatives *Ulothrix, Monostroma,* and *Ulva*, exemplifies the degree of relationship and of difference among closely related genera.

The *Enteromorpha* life cycle diagrammed here is typical for many marine algae. Haploid and diploid plants (see p. 31) are indistinguishable in gross structure, as are male and female plants.

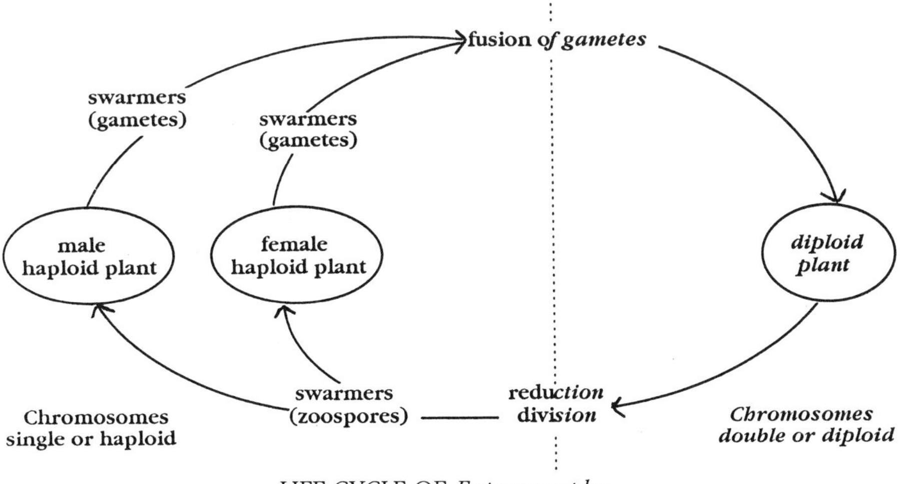

LIFE CYCLE OF *Enteromorpha*

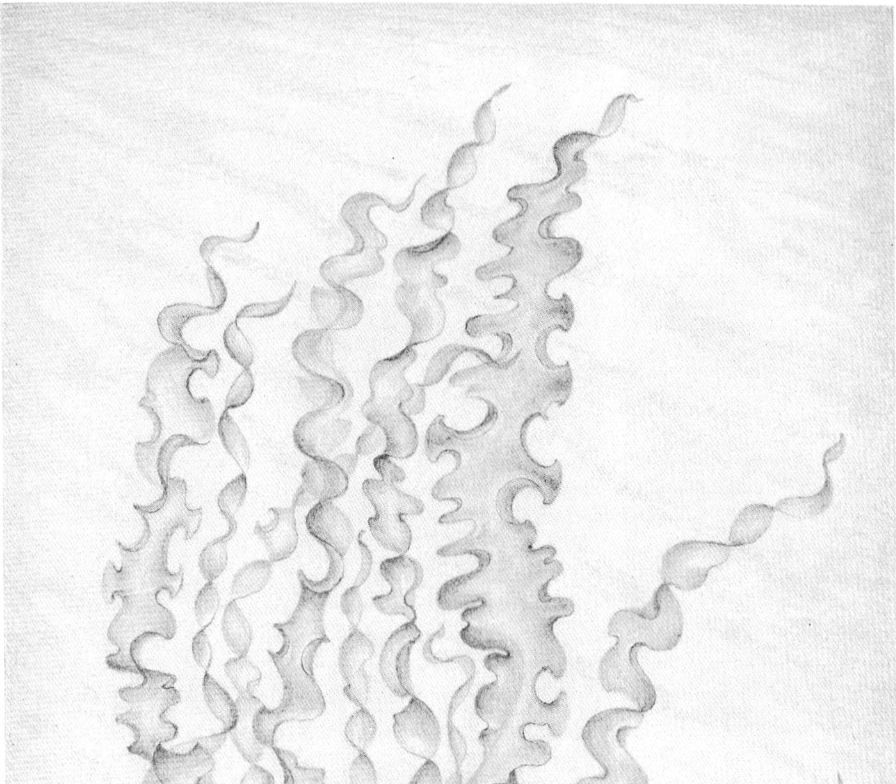

Enteromorpha linza X .50

Enteromorpha linza

Ulvales

Derivation: "gut-shaped."
Range: entire East Coast.

This *Enteromorpha*, widely distributed on rocks in exposed locations, is intermediate in structure between tubular species of *Enteromorpha* and the sheet-like *Ulva lactuca*. Each small, green blade is composed of a sheet of cells, two cells thick in the flat part, like an *Ulva*, but with each of the two edges and the stalk-like section at the bottom tubular like an *Enteromorpha*. A population of large, mature plants is shown here attached to a rock with a periwinkle nearby.

Some years ago, Kingsbury made a study of the windward and leeward faces of the long government jetty extending from the north shore across the mouth of the harbor at West Falmouth, in order to find out something about the effects of exposure to wave action on the composition of populations of seaweeds. The west, or windward face of the jetty, particularly at the tip, is one of the best places on Cape Cod to observe zonation. Here the rocks are placed to present a more or less regular surface which makes zones of growth easier to see, and the tidal change averages four feet in height, nearly double what it is along the south shore of the Cape and the Islands. This makes the zones broader. Zonation is a phenomenon that has excited considerable interest among biologists and has been well studied in such diverse locations as England, South Africa, and New Zealand. Zonation consists of the separation along a short vertical axis of populations of seaweeds and animals that are more or less homogeneous on the horizontal axis. Usually the major characteristics of zonation are obvious to the eye. Thus, at West Falmouth jetty, on the windward face, a spray zone with *Calothrix* is obvious as a black band above the level of average high tide. Next below comes a brown zone populated by *Fucus* (see p. 111). In mid summer, just below the *Fucus* zone is a bright green zone of *Enteromorpha linza*. Finally, a little above low tide line and extending downward comes a mixed population of red seaweeds, giving a band of red as deep as one can see into the water. Unlike the others, the *Enteromorpha* is seasonal. When it fruits and disappears in late summer, the rocks on which it grew are left almost bare.

Even in the zone of mixed red seaweeds, populations are not uniform. Close examination, by identifying every species and counting every individual in a given area, discloses a complex pattern of seaweed distribution which is not obvious to the casual observer.

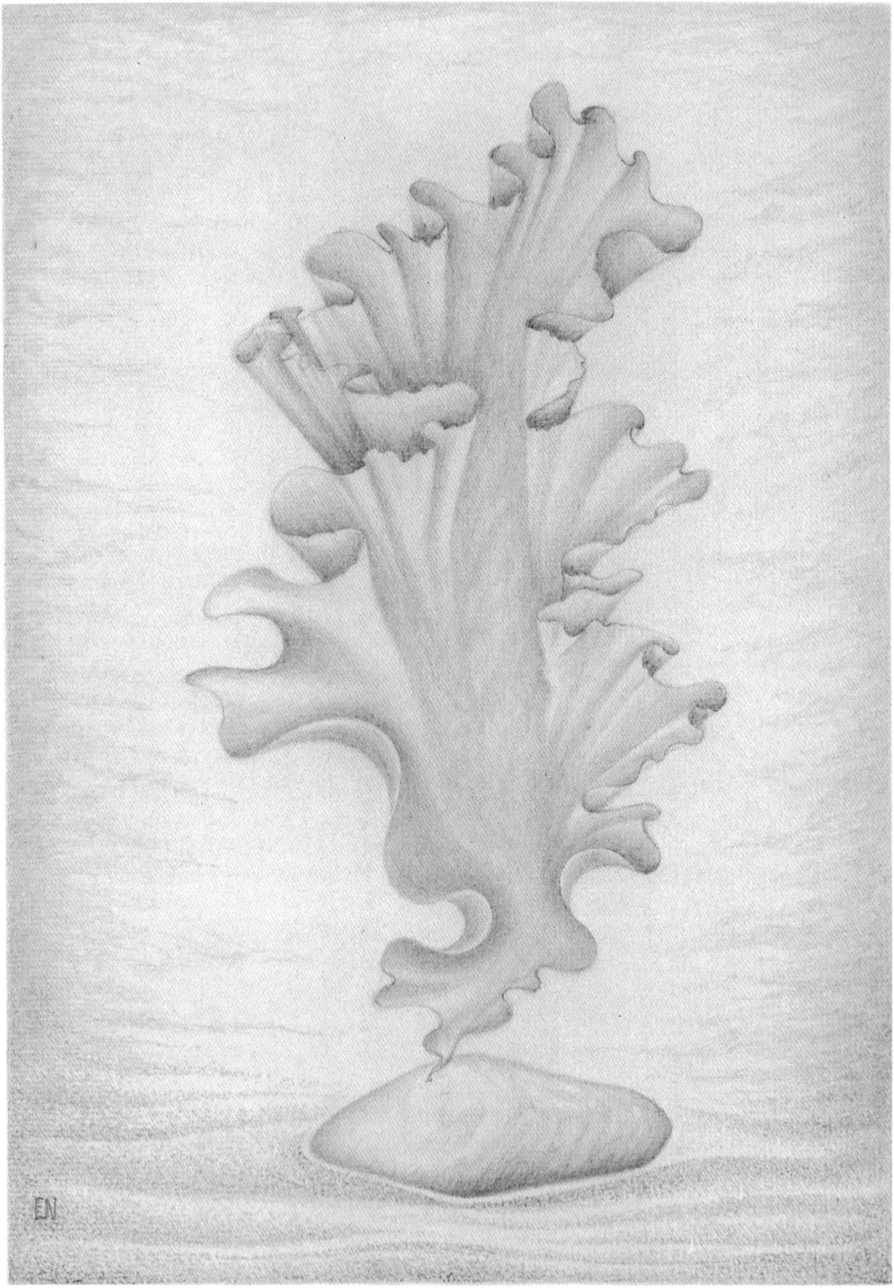

Ulva lactuca X 0.50

Ulva lactuca

Ulvales

Derivation: old Celtic word for water.
Range: entire East Coast.

This species name means "lettuce" and this plant is sometimes called "sea lettuce." It is very common between tide levels in moderately exposed locations. It also is found in warm, quiet waters where pieces of blade may be torn from the parent plant yet will continue growth, sometimes becoming a yard or more in length. Healthy, actively growing plants such as that depicted opposite are a silky-bright green. They have almost no stalk at the point of attachment. Older specimens become torn and perforated and may support the growth of small epiphytes.

Ulva may be confused with *Monostroma* or *Enteromorpha linza. Ulva lactuca* is usually larger than the latter and does not have the tubular edges (determined under a hand lens after sectioning them). It is coarser than *Monostroma* which consists of a sheet of cells only one cell thick. *Ulva* is two cells thick. A microscope is necessary to determine this characteristic for certain, but a good guess can be made on the basis of "feel." *Ulva* feels sturdier and thicker than *Monostroma*, something like the consistency of wax paper compared with tissue paper. You may see your fingerprints indistinctly through a single sheet of *Ulva*; you should be able to see your fingerprints clearly through *Monostroma. Ulva* is built like an *Enteromorpha* "ironed-out" so that the two layers of cells are stuck together back to back.

Ulva, and its relatives discussed here, are tolerant of wide variation in salinity and chemical composition of the water in which they grow. They are stimulated by the presence of additional nitrogen compounds. Nitrogen is scarce in ocean water. For these reasons, *Ulva* and its relatives are often found in particular abundance at locations of moderate pollution. State and federal laws are putting an end to the previously common practice of running a sewer pipe directly from the shore-side cottage onto the beach. But one can still find places where storm sewers, cesspool seepage, septic tank leakage, broken sewers, fertilized lawns, and agricultural runoff have increased the local nitrogen level and favored the growth of *Ulva*. Under enrichment of this sort, or in harbors and shallow, warm bays, *Ulva* may grow so luxuriantly as to form numerous massive blades. When these wash up on the beach and decay, people are unhappy.

Prasiola stipitata on bird rocks

Prasiola stipitata

Prasiolales

Derivation: "leek-green."
Range: Marthas Vineyard to Newfoundland.

This seaweed is unusual in habitat and nearly unique in life cycle. It forms dark, dirty green patches at or above high tide line on rocks where marine birds congregate, particularly on the vertical faces of the spray zone or in depressions that stay moist with spray. Here it is obviously favored by the fertilizing effect of bird droppings and can tolerate concentrations of nitrogen and other elements that might kill other seaweeds. Along most coasts the population of human beings is dense enough to keep gulls and other shore birds at a distance. They favor large, offshore rocks for roosting, and it is these "bird rocks" that often have dense coatings of *Prasiola*. It occurs, for example, on such rocks just off Gay Head on Marthas Vineyard, and on the long jetty at Sandwich by the Canal entrance.

A patch consists of a population of plants densely crowded together. They can be scraped off with a fingernail. The mass of green will separate itself in water into tiny green specks resembling flakes of parsley. Each is a separate plant composed of a sheet of cells one cell thick, as in *Monostroma*. But these cells differ from those of *Monostroma* in some internal characteristics. Each fleck, usually less than a quarter-inch long, narrows to the base and its point of attachment to the rock.

Some of these plants are diploid and reproduce themselves by non-swimming spores formed singly in cells at the outer edge of the blade. Others begin similarly as diploid plants. Then, individual cells scattered in the blade undergo cell division in such a way that the chromosomes are halved in the daughter cells (reduction division). The blade soon becomes a mosaic of haploid cells among the original diploid cells. Some of these will produce male gametes and others female gametes. The male gametes are motile swarmers like those of *Enteromorpha*, but the female gametes are non-motile, hence eggs. Ripe male and female gametes are shed from the haploid cells in which they are produced singly. They join and fuse in pairs, and the diploid fusion cell settles, attaches to the appropriate substrate and grows into a new diploid plant.

In having both phases of an *Enteromorpha* type of life cycle present and indistinguishable in a single plant blade, *Prasiola* is an unusual compound organism with no exact parallel among either plants or animals.

Chaetomorpha melagonium X 0.50

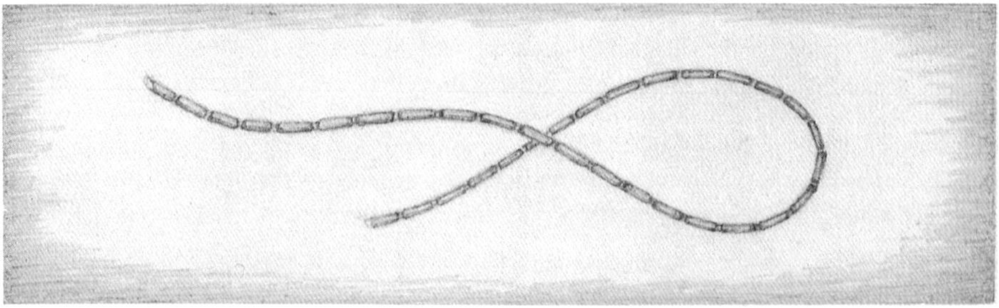

Chaetomorpha melagonium enlarged detail

Chaetomorpha melagonium — Cladophorales

Derivation: "hair-shaped."
Range: New Jersey northward.

The next few pages present the common representatives of a group of closely related filamentous green seaweeds of the Order Cladophorales. The cell structure among them is somewhat different from that described for the Ulotrichales (*Ulothrix* and relatives) and Ulvales (*Enteromorpha* and relatives), and in *Chaetomorpha melagonium*, the individual cells are so large (almost a sixteenth of an inch long) that they can be seen with the naked eye. All members of this group (Cladophorales) are filamentous in structure; each thread is composed of a single chain of barrel-shaped or cylindrical cells joined end to end. Some branch; others, including *Chaetomorpha* and *Rhizoclonium*, do not. Absence of branches can be determined with a hand lens, if not by eye.

Chaetomorpha melagonium is a cold-water seaweed. Farther north it grows abundantly on rocks in tide pools and below low tide line. On Cape Cod you may find it at such cold-water locations as the Sandwich jetty (mouth of the Canal), or at Gay Head on Marthas Vineyard which is close to an arm of deep, cold water from the southeast. Look for single, often entangled, unbranched, gray-green coarse strands, usually six inches to one foot long, growing in places where they are not exposed to drying.

Thirty cells of a large *Ulothrix flacca* laid side by side would be required to equal the width of a single cell of a large *Chaetomorpha melagonium*. These immense cells are useful to investigators because they can be entered with small instruments under the microscope without killing them, and various experiments can be performed that would be difficult or impossible in smaller cells. Inside the wall in each cell is a thick layer of living protoplasm. This, in turn, surrounds a vacuole or pocket of sap (non-living) in the center of the cell. Materials which accumulate or disappear from the sap vacuole must pass through the protoplasmic layer and the two membrane-faces that bound it.

Chaetomorpha linum x 1.0

Chaetomorpha linum Cladophorales

Derivation: "hair-shaped."
Range: entire East Coast.

Many filamentous green algae and some species of *Enteromorpha* form tangles of threads as they grow. The bright green curled tangles of *Chaetomorpha linum* can be recognized from others by virtue of their stiffness. When a tangle is removed from the water, it collapses only slightly in the hand, and the loops of individual filaments remain stiffly arched out of water. It looks (except for color) and feels like horsehair mattress stuffing or monofilament fishing line. A belief that green filaments of this sort have actually grown out of, or been transformed from horsehair was common in the days of public horse watering-troughs with their associated growths of filamentous algae. This species of *Chaetomorpha* grows on sandy and muddy bottoms in warm water. It is wide ranging, from Bermuda to Nova Scotia or farther. It starts attached to some solid object such as a small stone at first resembling *Chaetomorpha melagonium*, but it soon becomes free, and continues to grow, forming the characteristic tangles that drift ashore.

Most organisms are built up of cells. Among plants the cells are walled. Among green algae a principal substance giving stiffness to cell walls is cellulose. The submicroscopic strands of cellulose wind together like the fibers of a rope, and these fibers are netted together to form the substance of the wall. The orientation of the fibers can be investigated by examining the diffraction pattern they make when X-rays are passed through a cleaned wall. *Chaetomorpha* shows a pattern which indicates that the fibers are not laid down parallel to the surface of the wall (as would be expected) but are in two directions at almost precisely right angles to each other. The fibers themselves are inelastic. When a cell divides, the walls of the daughter cells must lengthen as each grows to adult size again. Somehow the fibers are loosened and slide on each other as the wall stretches. New fibers are added in this process.

The cell wall is dead and passively permeable, like a piece of filter paper. Thus the necessary functions of a living cell are carried out in the protoplasmic layer just within the cell wall. One of the major functions of protoplasm is the accumulation of certain needed chemical substances from seawater and the rejection of others. The composition of the sap in the center of a cell from a marine alga such as *Chaetomorpha* is usually quite different from that of the ocean water around it.

How some compounds are actively pulled into a cell from the surrounding water (or blood), while others are prevented from entering it, are matters of intense research. Answers could have great medical significance for humanity.

Cladophora sericea X 0.50

Cladophora sp. portion of plant X 1.0

Cladophora sericea Cladophorales

Derivation: "branch-bearer."
Range: Georgia to Labrador.

This genus is well named because the name emphasizes the major characteristic separating *Cladophora* from many of its relatives—the presence of branching. If you think you have a *Cladophora*, spread out a tuft of the fine green threads in a drop of water on your fingertips and look with the hand lens for branching. If the filaments are coarse, and especially if you can see the individual cells, and find no branches, you probably have a *Chaetomorpha*. If the filaments are fine and branching, it is probably either one of the many species of *Cladophora* or *Spongomorpha* (see the next description). If the filaments are fine, but you find no branches with the hand lens, you probably have *Rhizoclonium* which for all practical purposes is an unbranched *Cladophora*.

There are a number of species of *Cladophora* in the Cape Cod area. Distinguishing among them is difficult, requiring examination of characteristics such as cell size and arrangement of the branches. Some of the more delicate, filmy ones that form clouds of light green in the water must be distinguished from certain brown seaweeds (such as *Ectocarpus*) or red seaweeds (such as *Callithamnion)* of similar habit. This separation is relatively easy under a microscope, but without a microscope, you will have to depend on a careful look at the color. Most species of *Cladophora* are grass green like most of the green algae, but a few are an unusual gray-green or almost blue-green in color.

All species of *Cladophora* start out attached to some solid substrate. Some remain attached throughout their development, while others tear loose and continue to grow as floating masses. Both *Cladophora* and *Rhizoclonium* possess species that live in freshwater. A species of *Cladophora* is usually responsible for the coarse mats of filaments that form on farm ponds, fertile lakes, or untreated swimming pools in summer.

Rhizoclonium tortuosum is the only representative of this genus in our area. It is entirely unbranched and grows attached in tangles. When removed from water, it collapses completely, in contrast to the unbranched *Chaetomorpha linum* which is coarser and stiffer.

Spongomorpha arcta X 1.0

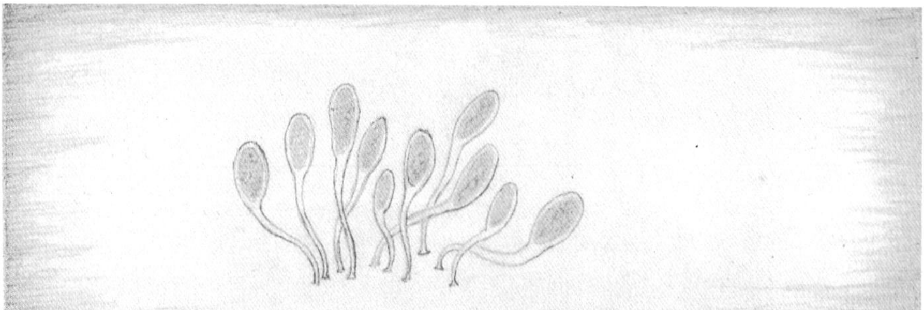

Codiolum gregarium greatly enlarged

Spongomorpha arcta — Acrosiphonales

Derivation: "sponge-shaped."
Range: New Jersey northward.

In structure, this is essentially a *Cladophora* with its filaments tangled together, the whole forming a definite, firm, wedge-shaped, spongy or ropey tuft of green. Often, the filaments have special, hooked branches. If you find such tufts, usually two or three inches long, hanging from rock surfaces between high and low tide lines of exposed locations in spring, or early summer, you probably have *Spongomorpha*. Another species, *S. aeruginosa*, forms somewhat smaller tufts. Both species are common farther north.

To determine the full life cycle of a seaweed usually requires getting it to grow and reproduce successfully under laboratory conditions. Sometimes this is difficult. When this was first done for *Spongomorpha*, its life cycle was surprising. At the time of reproduction, the protoplasm of each cell reorganizes and forms many small, motile swarmers. These are released through a pore that develops in the cell wall, swim about, and ultimately fuse in pairs with similar swarmers from another *Spongomorpha* plant. Thus, the swarmers are gametes. The fusion cell settles onto an appropriate substrate, but it does not grow into a diploid *Spongomorpha* filament as one might expect. Laboratory cultures showed instead that it grows into a plant which had previously been recognized as a species of *Codiolum*. Each *Codiolum* individual consists of a single large cell attached to the substrate at its lower end.

Mature *Codiolum* reproduces by swarmers (zoospores, produced by a reduction division) that grow into new plants of *Spongomorpha* filament. The life cycle is fundamentally the same as that for *Enteromorpha*, except that the alternating haploid and diploid plants are greatly dissimilar in appearance. To complicate matters, other green algae have diploid *Codiolum* stages. They include species of *Ulothrix* and possibly *Monostroma*.

We have found *Codiolum gregarium* on the exposed face of a rock on the southwest side of Penikese Island where it formed a dense population. Each very large, balloon-like cell is anchored to the rock by a long, colorless stalk. The population as a whole formed a dense green, somewhat fuzzy coating on the rock.

Bryopsis plumosa X 1.0

Bryopsis plumosa branch enlarged

Bryopsis plumosa Caulerpales

Derivation: "moss-like."
Range: entire East Coast to Newfoundland.

Bryopsis is constructed very differently from any other green seaweed in our range, but it looks, when taken out of the water, not unlike a small, loose *Spongomorpha* or *Cladophora*. It grows on rocks and wood in moderately protected locations at or below low tide. Under water it forms erect, bright green tufts, often two or three inches long. These appear to be composed of fairly large, branching filaments, but there are no cross walls, so that in reality the entire tuft is a single "cell". Well developed material displays a characteristic branching pattern. Each major axis is naked at the base, but toward the apex bears a number of branches in pinnate arrangement and of decreasing length. The frond is thus broadly triangular in outline like a stylized Christmas tree. Material should be examined under water to determine this characteristic.

Bryopsis and *Codium* (see next) are representatives of a group of green algae whose members are among the most numerous of the marine algae in tropical waters the world around. One unusual characteristic they share is that they do not form cross walls separating one cell from another. The protoplasm in each individual seaweed is continuous throughout. As the plant grows in length its nuclei divide, and thus the whole organism can be thought of as a single multinucleate cell. *Bryopsis* represents this tubular structural form at its simplest. In *Codium* the tubes become interwoven to form a large, complex three-dimensional structure. Individual *Codium* plants can each weigh several pounds.

On tropical reefs, relatives of *Bryopsis* and *Codium* are common and have evolved special adaptations to discourage herbivores. Some species form a hard outer covering of calcareous material, while others produce special chemical deterrents to feeding. Calcified green algae are a major source of the white sands of tropical beaches when waves break up dead individuals into small pieces, or parrotfish (which are not deterred by these algae) defecate the calcium grains that remain after digestion of organic materials from the plant.

Codium fragile X 0.50

Codium fragile

Caulerpales

Derivation: "fleece."
Range: North Carolina to Maine.

This seaweed is spectacular and is unlike any other. It consists of large, green or yellow-green, branching cylinders that feel like densely-packed, waterlogged felt. Branches may come out in "Y"s or irregularly. Mature plants may be almost a yard long and often weigh several pounds when wet. The major axes are about one-half inch in diameter. Each axis is composed of a web of colorless threads in the middle, running mostly lengthwise. These bear microscopic branches that turn outward, rebranch, and become green and swollen at the tips. The surface of the plant thus is composed of a tightly packed mass of swollen thread tips (called utricles) all projecting radially to about the same distance, giving the surface an appearance somewhat like the facets of an insect's compound eye. You can make this out with a good hand lens.

As in *Bryopsis*, all threads arise from a common source, and there are no cross walls. The protoplasm within the threads and ultricles is continuous throughout a healthy plant. Injuries to the seaweed that would otherwise cause its protoplasm to leak out are healed by a mechanism that creates new wall to repair the damage.

This species of *Codium*, the only one on Cape Cod, but one of some fifty found world-wide, is a weed and a nuisance. It readily attaches to shells, such as the scallop shell shown here, or small stones—and grows rapidly. It may interfere with the normal development of scallops or oysters, clogs clam rakes, and is conspicuous when it drifts ashore to die and decay on crowded beaches in summer. Prior to 1957 *Codium* had never been found on the western shores of the Atlantic. That year a few healthy plants were discovered at Orient Point, eastern Long Island. Since then *Codium* has gradually spread along our coast. Kingsbury found it at West Falmouth in 1967, and by 1968 it had reached the head of Buzzards Bay. Today there are isolated populations north of Cape Cod in the Gulf of Maine.

This invasion from Europe has clearly shown that distribution of seaweeds is controlled by barriers such as the open expanse and depths of the Atlantic Ocean. One can only speculate how the first *Codium* arrived here—possibly it was attached to the hull of an ocean-going ship or a fragment was carried in a ship's ballast. Well adapted as an invading species, *Codium* has the ability to reproduce readily by fragmentation rather than going through its full life cycle, and the fragments are easily distributed by the flow of water along our coast.

Ectocarpus siliculosus X 0.50

Ectocarpus siliculosus

Ectocarpales

Derivation: "fruit on the outside."
Range: entire East Coast.

This filmy, brown seaweed is common from Bermuda to Labrador and beyond, and is found attached to coarser seaweeds or other objects. Except for its color, it could be confused with a *Cladophora*. The basic branching filament structure is similar, but the color of an *Ectocarpus* is a true "middle" brown, instead of green, although sometimes *Ectocarpus* may have a yellowish cast. Dense growths may be almost black-brown.

Each individual plant, growing from a swarmer that has attached to some solid substrate, is an erect, branching and rebranching filament, composed of a single series of cylindrical cells joined end to end. With time, and a stable substrate, individuals of *Ectocarpus siliculosus* can become quite large despite their fundamentally loose filamentous structure. A mature specimen is shown opposite while younger individuals are shown in the second illustration (next page). Under water, the filments may separate widely giving the whole the appearance of a puffy cloud. Another species, *Ectocarpus fasciculatus*, is distinguished from *E. siliculosus* by differences in its plurilocular reproductive structures. The related genus *Spongonema tomentosum* has branches that are densely entangled into ropy strands, while the branches of *Ectocarpus* are more openly spreading.

The cells themselves in *Ectocarpus* are usually too small to be seen individually with the hand lens. Each is constructed much like the cell of an *Ulothrix*. It has a cell wall of cellulose within which is contained the protoplasm. Two major structures of the protoplasm are the chloroplast and the nucleus. Each cell contains only one nucleus, as is true for the cells of most brown algae. *Ectocarpus* cells contain only one band or plate-shaped chloroplast each. Related filamentous brown algae with more than one chloroplast per cell are placed in the genus *Hincksia* of which there are several species, but to see the chloroplasts requires viewing with a good microscope. Brown algae owe their characteristic color to a larger concentration of certain carotenoid pigments in the chloroplasts, in proportion to the chlorophyll, than is found among the photosynthetic pigments of green algae.

The swarmers of brown algae, whether gametes (fusing sexually) or zoospores (direct-germinating), are microscopic, flagellated bodies that swim rapidly in sea-water. The swarmers in most algal groups are egg-shaped with two or four thread-like protoplasmic extensions (flagella) inserted at one end. In the brown algae, swarmers have a unique and characteristic shape, something like that of a kidney bean, with the two whip-like flagella inserted laterally at the point where the scar would be on a bean. Swarmers are not produced in the vegetative cells of the filaments as in most green algae, but in specialized reproductive structures, the single-celled sporangia and the multicelled gametangia or sporangia. These structures are described under *Pilayella*.

The life cycle of *Ectocarpus siliculosus* was investigated in the early 1930s at Woods Hole on Cape Cod in pioneering work by Dr. George Papenfuss. What he discovered has proved a major contribution to knowledge of life cycles among brown algae. *Ectocarpus siliculosus* is easily collected in summer attached to *Chorda*, to other coarse algae, and to other objects throughout Cape Cod waters. These plants, Dr. Papenfuss found, bore reproductive structures (plurilocular) in which swarmers were produced that were neither the progeny of a reductive division, nor did they fuse as gametes. Instead, they grew directly into more plants exactly like the parent, and any potential cycle between haploid and diploid plants was thus short-circuited. Later in the summer another type of reproductive structure (unilocular) was found on the *Ectocarpus* plants. Its swarmers did not fuse as gametes but instead also germinated directly. It was not possible then to follow further development through to the next reproductive phase. Finally, on a trip with the botany class from the Marine Biological Laboratory at Woods Hole to Penikese Island at the very end of the summer, Dr. Papenfuss collected and preserved some *Ectocarpus siliculosus* growing on another brown seaweed, *Chordaria*. When he examined these specimens the following winter the number of chromosomes in each vegetative cell (haploid) suggested that he probably had the missing sexual stage in the life cycle. Where did they come from?

Penikese Island, a small dot of land in Buzzards Bay, was then a remote bird sanctuary. Dr. Papenfuss tried the next summer to make collections again from the same place but could not find sexual (haploid) plants at that time. Finally, in November after several further abortive attempts to collect sexual material, he arranged to spend time in the cottage of the island caretaker, the only person then on the island, and he then found sexual plants a second time. This time he watched them release swarmers from plurilocular structures, and watched the swarmers fuse, proving their gametic nature.

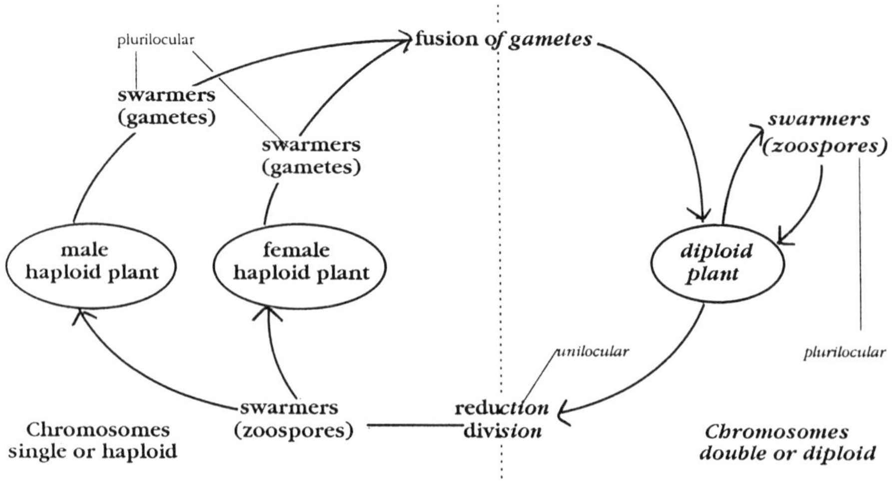

LIFE CYCLE OF *Ectocarpus*

Ectocarpus siliculosus on *Chorda* X 1.0

Pilayella littoralis on *Fucus* X 1.0

Pilayella littoralis

Ectocarpales

Derivation: honors de la Pylaie, a French botanist.
Range: Virginia northward.

It is often impossible to distinguish this plant from species of *Ectocarpus* by the naked eye or a hand lens. *Pilayella littoralis* grows in similar habitats (here, epiphytic on a *Fucus* plant) and looks much the same as *Ectocarpus siliculosus.* There are fundamental differences, however, in the structure of the filament, and especially in the structure of the reproductive bodies.

As described on the preceding page, *Ectocarpus siliculosus* produces at least two different types of reproductive bodies. They are formed laterally as short branches from the axis bearing them. One type, the unilocular ("one-cavitied") looks something like a lightbulb. It starts out as a single cell with one nucleus. Then the nucleus divides repeatedly and the protoplasm in the "lightbulb" separates itself into numerous swarmers of characteristic shape which form two flagella apiece, and are released. The other type of reproductive structure is much different in appearance, although it, too, is produced as a lateral branch. This structure is larger than the unilocular, looks something like a baseball bat attached at the large end and is divided into a great number of very small cells, each with its own cell wall. Inside each cell of this plurilocular ("many-cavitied") structure the protoplasm reorganizes about the single nucleus and forms a single swarmer identical in appearance to the swarmers produced in the unilocular structure.

Among brown algae, it turns out that reduction division (of chromosomes) always takes place in the unilocular structure, which is produced only by the diploid plant, and its swarmers are always zoospores (direct germinating). The plants they grow into are haploid, sexual plants. If the plurilocular structures are borne on diploid plants, their swarmers are diploid zoospores and reproduce the diploid stage directly, short-circuiting the sexual stage. Haploid, sexual plants, grown from the swarmers from the unilocular structures, produce only plurilocular structures at the time of fruiting. This is the kind of *Ectocarpus siliculosus* that Dr. Papenfuss found on *Chordaria* at Penikese. The swarmers from these plants are haploid, and fuse. Therefore they are gametes. The diploid fusion cell grows into the diploid stage and the cycle is completed.

Pilayella has both unilocular and plurilocular reproductive structures, but instead of being produced at branch tips, they are produced along the length of a regular filament, the unilocular structures in a series. Also, some vegetative cells of the major axis in *Pilayella* may develop an occasional longitudinal wall so that at these locations the filament is two cells wide.

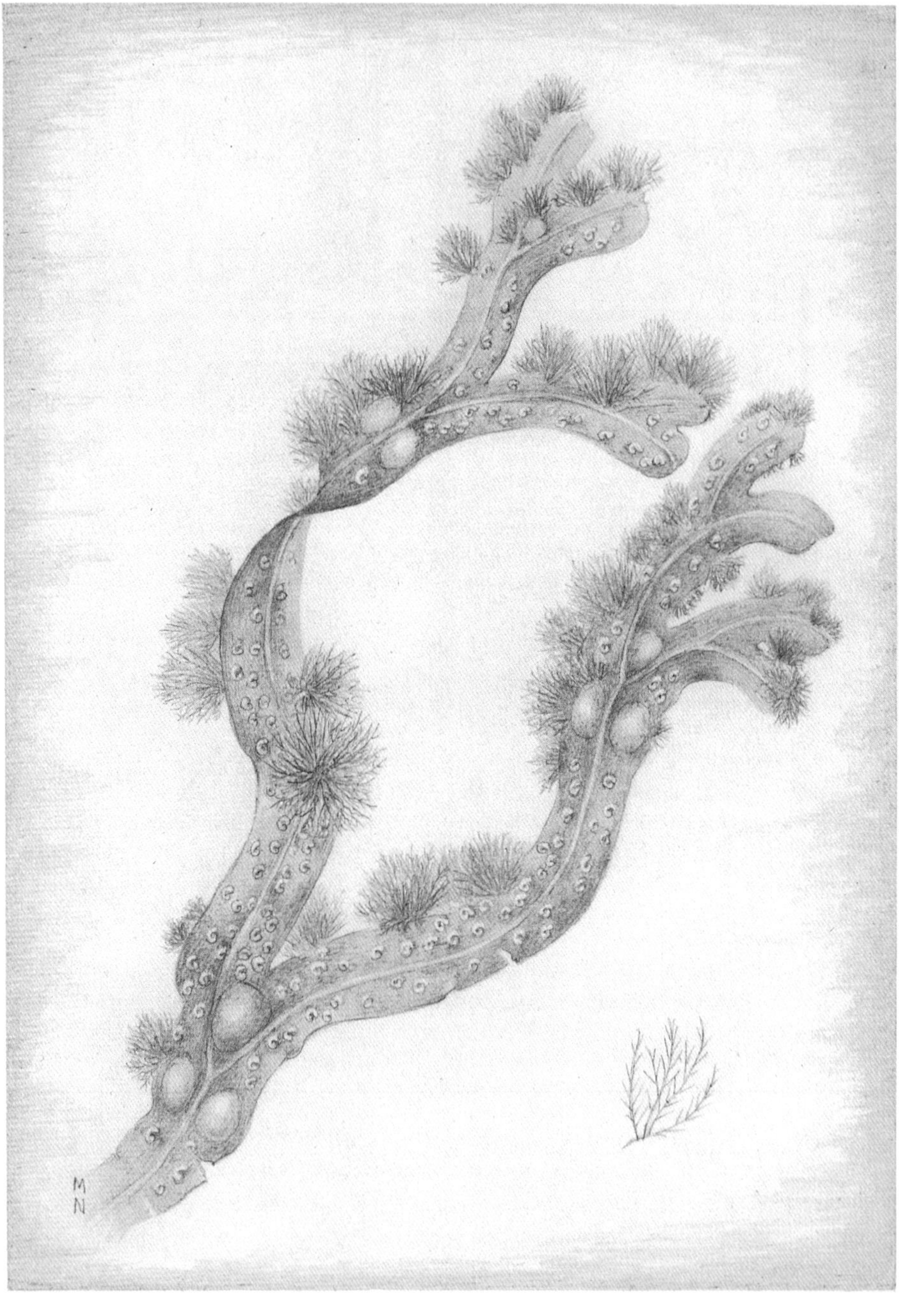

Sphacelaria cirrosa on *Fucus* X1.0 branch detail

Sphacelaria cirrosa

Sphacelariales

Derivation: "gangrenous."
Range: Virginia northward.

Ectocarpus, a primitive brown alga, displays diffuse growth—cells of the filament divide randomly, without obvious pattern. In *Sphacelaria,* which appears filamentous to the naked eye, cell divisions contributing to growth in length occur repeatedly in a single large cell at the tip of each axis or branch. These apical cells are somewhat swollen, dense of contents, and much darker brown than the cells below. When dry, as in a preserved specimen, they become shrivelled and black, and it is in reference to this characteristic that the name "gangrenous" was first applied. In German they are called "Brenntalgen" ("burned algae") for the same reason.

This species, the commonest of several that occur in Cape Cod waters, forms hemispherical tufts, each about a half inch in height. It is frequently found growing on the blades of *Fucus,* but occasionally may be found (usually less well developed) on other coarse seaweeds or shells. Several other algae form small tufts on *Fucus. Sphacelaria* tufts can be suspected if they are stiff, not relaxing when taken from the water. A more certain characteristic for identifying them is the fact that the branch tips often are rebranched in pinnate fashion (like the structure of a feather, or of most fern fronds) when viewed with a good hand lens. Unfortunately, not all specimens show good pinnate structure for one reason or another.

Although they appear so to the naked eye, the axes of a *Sphacelaria* are not simple filaments. Additional longitudinal divisions take place in each cell of the axis, but without causing an increase in length or diameter of the axis. The appearance of a matured axis, as seen under the microscope is not unlike an elongated Tower of Pisa. This type of structure, in which the original cells of an axis divide secondarily in longitudinal fashion is termed polysiphonous.

Reproduction in *Sphacelaria* is fundamentally similar to that in *Ectocarpus* and *Pilayella,* but is a rare event. Instead, a special type of vegetative propagation is common. Small specialized lateral branches are produced on the main axis. When fully developed, each has two or three arms almost at right angles, and looks something like a child's jack. The point of attachment to the axis is very narrow and breaks easily. When a drifting propagule, as this structure is called, falls on an appropriate substratum, it attaches and begins to grow forth as a new plant.

Cladostephus spongiosus X 0.60

Cladostephus spongiosus enlarged detail

Cladostephus spongiosus Sphacelariales

Derivation: "crown of branches."
Range: Long Island to New Hampshire.

Most of the members of this Order of seaweeds (Sphacelariales) are found in marine waters of the southern hemisphere. We are lucky to have the two particular common representatives that we find at the latitude of Cape Cod because one *(Sphacelaria)* represents a simple type, while the other *(Cladostephus)* is typical of more complex members of this group. The branching plants of *Cladostephus* are sufficiently different from any other seaweed that recognizing it should present no problem. Close examination by naked eye, and especially with a hand lens, shows that each major axis is beset with densely packed whorls of branchlets, and the plant has much the structure of a bottle brush.

Cladostephus attains this complex structure essentially by elaboration on the *Sphacelaria* theme. It starts out with the same type of dividing apical cell and polysiphonous construction, and then continues growth. Each major axis undergoes further divisions and increases in size, especially diameter. No one would confuse it with a simple filament. Unlike *Sphacelaria*, *Cladostephus* has no specialized structures for vegetative propagation.

Cladostephus is a seaweed of warm, relatively deep water, where it lives attached to stones. You are unlikely to find it attached without either diving or dredging for it. On the other hand, good specimens are frequently found in the wash. The stiff, wiry branches persist for a long time before decaying. A good place to look for it would be along the beaches of the south shore of the Cape after a storm has stirred the waters of Nantucket and Vineyard Sounds.

The illustration depicts a single plant attached to a stone with the shell of an oyster drill in the background. The closeup shows how the whorls of branchlets are borne on the major axes.

Fruiting in *Cladostephus* is seasonally initiated. The axes shed their branchlets and die back more or less. New simpler branchlets appear irregularly from the old surfaces. Unilocular or plurilocular reproductive structures (depending on stage in the life cycle) are formed on these; swarmers are produced and shed. Swarmers are typical of those found among brown algae. Each bears two laterally inserted flagella which give it motility. Motility of swarmers, whether of green or brown algae, is not random. Usually swarmers display a distinct orientation in regard to light, swimming either strongly toward it or away from it. The structure within the swarmer which helps the cell respond to the direction of light can be seen under the microscope as a tiny orange dot associated with a chloroplast. It is often called an eyespot. The eyespot shades a photosensitive part of the flagellum, which orients the cell to the light.

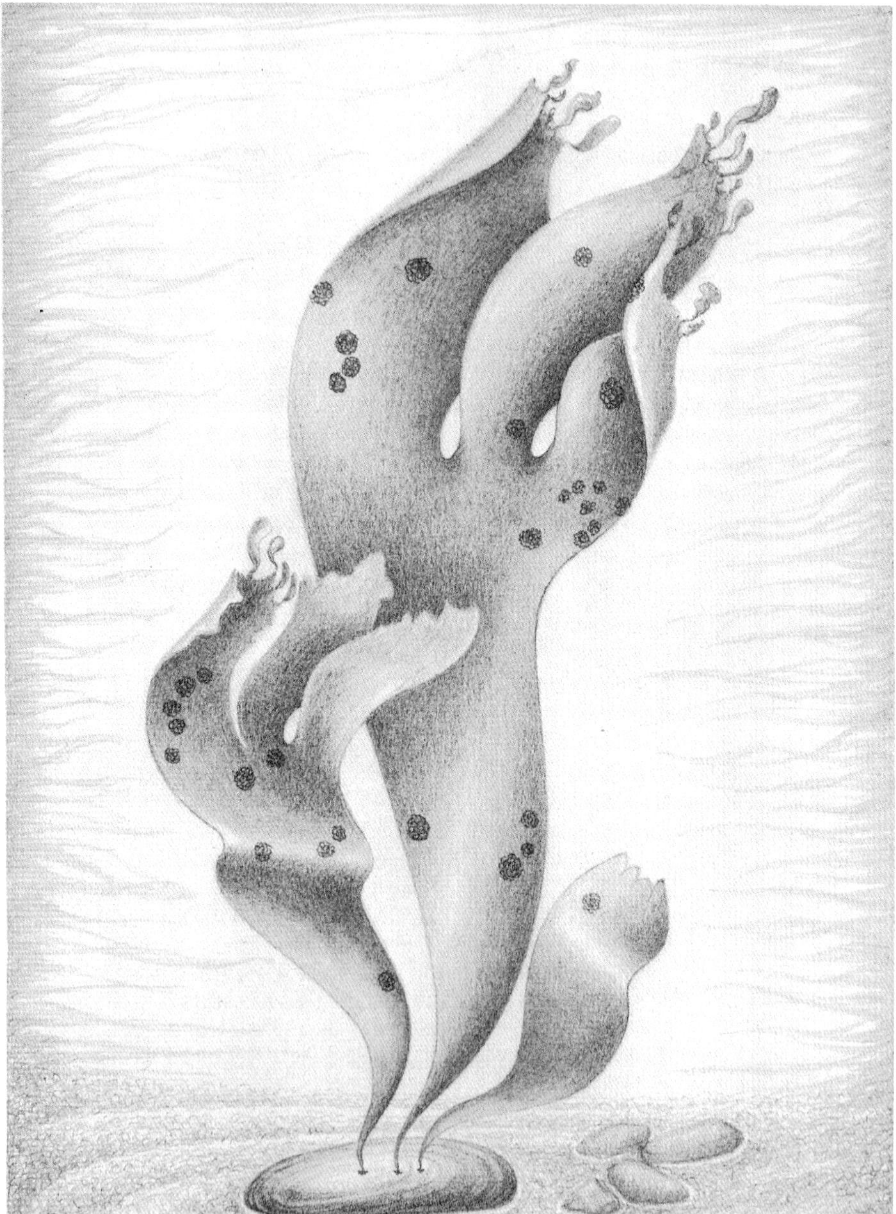

Myrionema strangulans on *Palmaria* X 1.0

Myrionema strangulans

Chordariales

Derivation: "numerous threads."
Range: entire East Coast.

This alga is included as an example of an inconspicuous, easily overlooked sort. There are many others like it, mostly microscopic or nearly so, that are beyond the general scope of this book. To find *Myrionema strangulans*, first find old *Ulva* or *Palmaria* on which it commonly grows. Then look for small, brown or discolored spots, irregularly circular in outline. Lots of things, such as the rasping of a snail, the death of a patch of cells after injury by a stone, or localized swarmer formation and discharge can produce such spots. Thus it will be difficult to determine with only a hand lens that any particular spot is composed of a living, filamentous plant.

Myrionema begins growth when a swarmer successfully settles and attaches to an appropriate host plant. Attachment here, as in most cases when swarmers settle, is by means of a zone or pad of mucilaginous wall material formed between the swarmer and the substrate. Of course, the swarmer loses its swimming flagella during this process if not before. After attachment, the original cell swells and divides. In many simple algae, as *Ulothrix*, the daughter cell is cut off upwards and the erect filament is thus initiated with a holdfast cell at the bottom. In *Myrionema*, however, the initial cross wall divides the original cell into daughter cells that both lie on the surface of the substrate. With further divisions, the original cell serves as the initiator of a number of radiating and branching filaments which remain prostrate against the substrate as they grow. Branching is so frequent that the whole forms a single layer of more or less tightly packed filaments (whence the name). The layer thickens as few-celled erect filaments grow upward from the cells of the original prostrate filaments to form the mature plant.

The same type of construction can be found in certain green and red algae, as well as here in the browns. Thus, when looking for *Myrionema*, you may find instead a distinctly red or green spot. Nor is *Myrionema* the only brown seaweed to show an encrusting, epiphytic type of growth. Some forms remain entirely pressed against the host with no erect filament branches at all. Others are more extensive.

Myrionema is the first member of the Order Chordariales, and is one of its smaller and simpler respresentatives. This Order of brown algae is distinguished from the preceding algae by specific differences in its life cycle and its plant structure, as described on page 67.

Elachista fucicola on *Fucus* X 1.0

Elachista fucicola

Chordariales

Derivation: "very small."
Range: Virginia northward.

The species name means "*Fucus*-loving" and on *Fucus* (or *Ascophyllum*) you should look to find this alga. Here it is shown on a plant of *Fucus vesiculosus*. Old fronds of *Fucus* often look bearded. If the "beard" occurs in discrete clumps (which often tend to run together), and particularly if the clumps of whiskers are very dark brown and have a hard knob at the base, you probably have a mature specimen of *Elachista fucicola* . Young plants are less likely to have developed the firm basal structure which feels like the head of a small hat pin when rolled between the fingers.

Elachista starts growth like a *Myrionema*. Young prostrate stages are followed by the development of several types of erect filaments. The long, deeply colored free filaments can be seen by naked eye or hand lens, but the other types of short filaments packed tightly together at the base of the long filaments cannot be distinguished without a microscope. It is the mass of short filaments that constitutes the "knob" you feel. *Elachista* is considered advanced over algae like *Myrionema* because it possesses the beginnings of differentiated "tissues." This is shown by the separation of specific functions into specific parts of the plant. The photosynthetic function is relegated mainly to the long, free filaments, whereas the other, tightly packed filaments in the center of the plant are nearly colorless. They become involved, instead, in reproduction at the proper season.

Elachista must be distinguished from young tufts of *Ectocarpus* or of other brown and red algae that may grow attached to *Fucus* and other coarse algae. On land, competition for living space seems to be less severe than at the fringes of the ocean. Almost all seaweeds require a place for attachment in order to exist, and "sitting-down-places" are scarce. It is not uncommon for three or even more unrelated seaweeds to be attached to each other sequentially, *a* on *b*, *b* on *c*, and *c* to a rock. One also often finds three or four unrelated algae growing on another single plant which in turn is attached to a rock (*a* & *b* & *c* on *d* on rock). Plants that grow on other plants are called epiphytes, and the plant which is grown upon may be termed the "host". Relationship of epiphytes to hosts is varied, interesting, and sometimes quite specific or subtle, as is explained on page 207. *Elachista fucicola* grows epiphytically on a few genera other than *Fucus*, but not many.

Leathesia difformis on *Chondrus* X 1.0

Leathesia difformis

Chordariales

Derivation: honors G. R. Leathes, a British naturalist.
Range: North Carolina northward.

Leathesia usually looks like oversized, waterlogged, brownish-yellow popcorn. The plants shown here on *Chondrus* are young. Older plants are more convoluted and irregular in outline. There is no other seaweed like it on the East Coast. It starts growth as a small lump on some other alga, commonly *Chondrus* or *Corallina*. At this stage its microscopic structure is not unlike *Elachista* but more branched. Its final branches terminate at the surface in a compact layer. *Leathesia* grows more or less equally in all directions, forming a larger and larger lump, with a cortex of radiating filaments on the outside and medulla of mostly randomly oriented filaments within. As further growth takes place, the medulla breaks down or is torn apart so that a medium-sized or larger plant is hollow and irregularly lumpy. If the plant starts on a small seaweed, it may come to engulf the latter completely. Plants need light, so a completely embedded host plant may be expected to die if the *Leathesia* about it gets too dense. Then the *Leathesia* may appear to be growing directly on the rock itself.

Leathesia difformis (our only species) appears in summer, fruits in autumn, then disappears. It overwinters in the form of a microscopic phase of the life cycle.

The Order Ectocarpales groups together the simple, mostly filamentous brown algae which possess two equal and indistinguishable phases in the life cycle. The Sphacelariales, though similar in life cycle, are set off on the basis of the rather unusual and definite structural characteristics shared among them. The Order Chordariales includes brown seaweeds built on a plan of interwoven filaments, loosely to densely packed according to the particular species and differentiated into cortex and medulla. This group, *Myrionema* through *Acrothrix*, is further united on the basis of a life cycle in which the plant as collected alternates with a microscopic gamete-bearing phase.

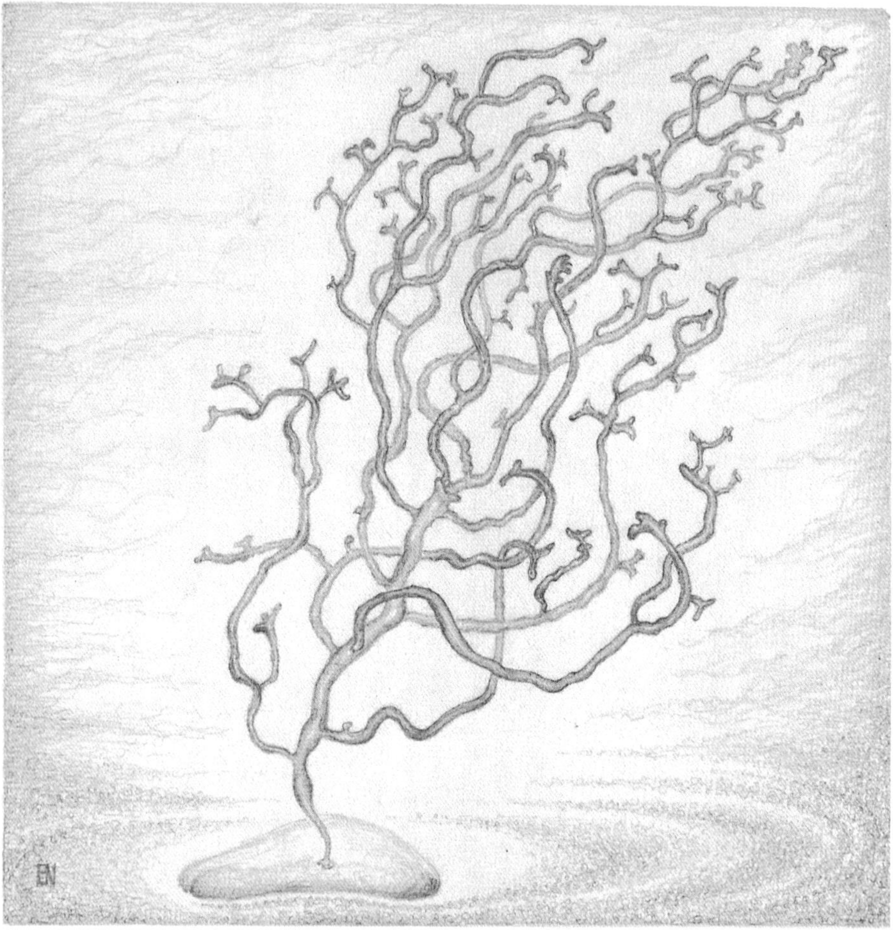

Eudesme virescens X 0.5

Eudesme virescens

Chordariales

Derivation: "well-bundled."
Range: Long Island northward.

The unusual consistency of *Eudesme* helps to identify it. The branching, soft, pale brown, gelatinous columns up to a foot or more in length are like slightly gone-by worms, or rather thick eggwhite, and are correspondingly difficult to handle out of water. In water the major axes are irregular in dimension, somewhat indefinite in outline, and coarsely branched. The branches may branch again irregularly, usually at a wide angle, and usually near the tips. When dried on paper, this plant seems to leave hardly more than a shadow. Typically, it lives attached to rocks, pilings, and occasionally to large algae near or below low tide line, and appears in greatest abundance in warm water locations in spring.

Eudesme may not create an appealing picture from the description above, but it has great value in teaching. It displays well one of the major types of structure found among seaweeds. Because of its softness, the microscopic structure can be shown easily simply by pressing a piece of the plant gently under the cover glass of a microscope slide. Few other brown algae of similar construction are soft enough to handle this way.

Even with just a hand lens, using material pressed out between glass, it should be possible to observe that the central part of the axis is nearly colorless. It consists of a loosely interwoven tangle of large-celled filaments or threads running length-wise. The outer part of each axis is dark-colored and is made up of a forest of radially oriented short filaments, all of which terminate at about the same level, and the tips of which, therefore, become the outer surface of the plant. This is not unlike the basic structure of *Codium,* except that in *Eudesme* the plant is built of filaments composed of discrete cells end to end in contrast to the single-cavitied threads of *Codium.* The radial filaments of *Eudesme* arise as branching systems from the ones running longitudinally and constitute a regularly distributed, furry covering around the latter. They are darker in color because their cells possess chloroplasts. Chloroplasts are largely lacking in the cells of the longitudinal central filaments. When a larger alga has structure more or less separable into distinct areas or zones on the basis of color, as here, the colored zone (outside) is termed the cortex, and the colorless zone (inside), the medulla. Photosynthesis is accomplished primarily in the cortex.

Sphaerotrichia divaricata on *Fucus* X 0.75

Sphaerotrichia divaricata

Chordariales

Derivation: "spherical filament."
Range: New Jersey northward.

This plant when actively growing is slippery like a wet watermelon seed. Unlike *Eudesme* it has palpable substance, and the slipperiness is the result of copious slime secreted by the cells at the surface. Another useful clue for recognizing *Sphaerotrichia* is the branching pattern, particularly the usual presence of at least a few short, widely angled branches or branch tips shown in great abundance in this illustration. If you find a brown seaweed with these characteristics, you will almost certainly have *Sphaerotrichia divaricata*. Unfortunately, in its form *Sphaerotrichia* is like many other brown seaweeds. Old plants lose their slipperiness, and under some conditions specimens may not display the characteristic branching pattern. In such cases, only an examination of cellular structure with a microscope will serve to identify it securely.

Sphaerotrichia is usually a "middle" brown or slightly yellow-brown in color. It forms conspicuous masses typically attached to various coarse seaweeds near or below low tide, especially in coves, bays, and other places of quiet waters.

In structure, *Sphaerotrichia* is similar to *Eudesme*, with a central medulla of tangled longitudinal filaments and a cortex of radially directed filament-branches. Compared with *Eudesme*, the filamentous structure of this alga is more firmly knit together, but compared with other seaweeds, it is still relatively loose. Young branch tips can usually be pressed out adequately for examination under a cover slip on a glass slide. Older parts become firmer. This genus takes its name from the fact that the terminal cell of each radial cortical filament is greatly swollen and spherical. The plant's surface is thus composed of a single layer of such cells.

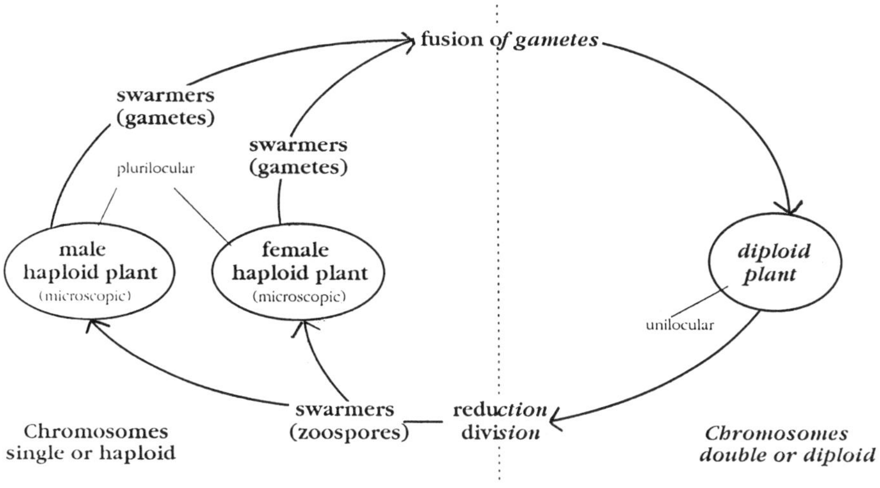

LIFE CYCLE OF A MEMBER OF THE ORDER CHORDARIALES (see p. 73)

Chordaria flagelliformis X 0.45

Chordaria flagelliformis Chordariales

Derivation: "cord-like."
Range: New Jersey northward.

Chordaria is usually about as black as a brown alga gets. It is a very common plant from Cape Cod northward, growing on rocks or wood in moderately exposed locations near or below low tide. Rarely do you find a single plant. If it is there at all it is usually present in abundance. As the summer season progresses it tends to accumulate epiphytes such as the brown *Ectocarpus* and various delicate red algae such as *Callithamnion*, and its general appearance becomes modified. Typical, well-developed plants of *Chordaria* (we have only one species) possess a short but obvious main axis. This is attached to the substratum and bears, in turn, the long, whip-like branches. The upper branches from the main axis are longer than the main axis, and are unbranched, straight-flexuous, and of regular diameter except tapering toward the tip. Along the lower part of the main axis the branches may be short and stubby as shown here. The whole plant is very firm, cartilaginous and elastic in texture, and may be somewhat slippery. A typical plant is easy to recognize by these characters.

Chordaria is similar in basic structure to *Eudesme* and *Sphaerotrichia* but the filaments of its cortex and medulla are pressed together very firmly indeed.

The typical pattern of reproduction in members of the Order Chordariales is diagrammed on p. 71. The plant which you find releases swarmers from unilocular structures similar to those described for *Ectocarpus* (page 54) when it reaches reproductive maturity. These swarmers result from a reductive division. The specimen you collected was thus diploid and its swarmers are haploid zoospores. As in *Ectocarpus siliculosus*, haploid zoospores grow into haploid plants which produce gametes at reproductive maturity restoring the diploid condition after sexual recombination of genetic material. The life cycle diagram is essentially the same as that for *Ectocarpus*. However, the gamete-producing plants of typical members of the Order Chordariales, and of most other brown algae, are microscopic, filamentous plantlets which in nature would be entirely overlooked. Our knowledge of the haploid, gamete-bearing plantlets comes entirely from cultural studies in which reproductive cells have been made to germinate and grow under laboratory conditions where they can be watched from day to day. The gametes are formed and released from plurilocular structures as described for *Ectocarpus*. The diploid cell which results from the fusion of gametes settles, attaches, and grows into a new diploid plant.

Chordaria flagelliformis and a number of other members of the Chordariales do not always follow the typical pattern but "short-circuit" the life cycle described above by eliminating the sexual stage.

Acrothrix gracilis X 0.45

Acrothrix gracilis

Chordariales

Derivation: "hair-tipped."
Range: Long Island to Newfoundland.

The first scientific publication to deal with New England seaweeds was a paper appearing in 1847 by Jacob Whitman Bailey, Professor at the West Point Military Academy. By 1880, observations had progressed to the point where Professor W. G. Farlow, of Harvard felt able to state about our seaweeds that "probably all the large and striking species are now known..." In 1928, Dr. W. R. Taylor noted that "more than fifty years have passed since this opinion was expressed, and it can hardly be said to have been without justification." Yet in that same year Dr. Taylor reported for the first time the appearance of *Bonnemaisonia* and *Acrothrix* on our shores. Both were well-known European plants probably introduced in similar fashion to the introduction of *Codium* more recently (page 51). Had *Acrothrix* been here and overlooked all these years? Unlikely, especially since the material Dr. Taylor described had been found up to the time of his report only in two isolated areas of Cape Cod (outer shore of Penikese Island, and Nobska Point). At the former it had been collected each year for three years; at the latter it was first collected the year before the report was written. Now it is well distributed and has been collected repeatedly from Long Island throughout southern Massachusetts, and elsewhere.

Acrothrix is another of those brown seaweeds of branching cylindrical habit, various species of which are often difficult to distinguish. This plant is moderate-sized (usually less than a foot tall), open-bushy, broadly spreading in the water, and clearly brown in color. Close examination will show one or more short main axes spreading widely from near the holdfast. These bear numerous long, curved, major branches which exceed the main axis in length. The widest axes are not more than 1/16th inch in diameter. *Acrothrix* can be differentiated from *Dictyosiphon* by its branching pattern (the latter has a main axis as long as the plant). It differs from *Chordaria* in its lighter brown color, and in the branching pattern. Unlike *Chordaria*, *Acrothrix* has a large number of tertiary (or third order) and higher order branches. For this reason *Acrothrix* usually looks bushier than *Chordaria*. To determine the order of branching, count succeeding branch points. The main axis (A) bears a number of first order branches (B); each of these bears second order branches (C); and so forth. *Acrothrix* grows on rocks, coarse seaweeds, and other objects in the protected, shallow waters of Buzzards Bay, Vineyard Sound, and Long Island, fruiting in late spring and early summer.

Desmarestia viridis X 0.10

Desmarestia viridis

Desmarestiales

Derivation: honors A. G. Desmarest, a French naturalist.
Range: New Jersey northward.

This is perhaps the only seaweed on Cape Cod that can usually be recognized by smell; but its particular acrid odor is difficult to describe. The plant is a large, brown to olive, soft, bushy, much branched and rebranched seaweed, often considerably more than a foot in extent as it grows attached to rocks just below low tide and deeper. The basic pattern of branching is opposite, but this may be obscured, particularly in the outer parts of the plant. The main branches are round, about 1/8 inch in diameter, and not slippery. The plant is an annual, appearing luxuriantly in early spring. At that time it is rebranched three to five or more times and the ultimate branchlets consist of filaments (single series of cells). The filamentous branchlets are shed as the season progresses, but even then the smallest remaining branches still are much less than a sixteenth of an inch in diameter. Points on the axis where branches originate may be slightly flattened.

This peculiar odor relates to the high acidity of the cell sap. As long as the plant remains alive it manages in some unknown way to keep its sap under control. As soon as it dies, however, the sap begins to dissolve or digest the cells which contain it. Within a very brief time, a healthy plant will be discolored green and will decay rapidly, especially if kept in a bucket with a small amount of sun-warmed water. Under these conditions the odor is particularly noticeable. The acidity of of the sap is caused by the ability of the cells to accumulate remarkable concentrations of sulfate from seawater, forming what is essentially a strong solution of sulfuric acid within the cell. It is the presence of this acid and its effect on dead seaweed which produces the odor. The specific name "viridis" comes from the Latin for "green" and reflects the tendency for *Desmarestia* to turn this color as it begins to decay.

Desmarestia viridis is difficult to bring home alive, and dangerous in the collecting bucket. It can destroy other delicate seaweeds in the bucket before you get them home. If you wish to make a good preserved specimen of it, you should collect it living (not in the wash), get it home, pressed, and dried as soon as possible. The sulfuric acidity of some species of *Desmarestia* is so high that a chemical reaction between the plant and the mounting paper creates a deep blue stain in the paper.

Sulfate is very dilute in the oceans, but *Desmarestia* concentrates it many thousand fold and is able to contain the acid in its sap without injuring the adjacent protoplasm. The high acidity may serve to deter feeding by herbivorous snails and sea urchins.

Desmarestia aculaeata X 0.45

Desmarestia aculeata Desmarestiales

Derivation: honors A. G. Desmarest, a French naturalist.
Range: New Jersey northward.

Compared with *Desmarestia viridis* and with other similar seaweeds, *Desmarestia aculeata* is stiff, almost like broom bristles. Its appearance in summer is very different from its early-season appearance when it is densely covered with small clusters of fine, filamentous hairs which are shed in early summer. Plants are relatively large, and branched and rebranched several times. The branches are quite conspicuously flattened. The edges of each branch typically bear sharp, forward-pointing teeth at regular intervals in alternate, pinnate arrangement. Less commonly found is a variety, probably from deep water, in which the branches are greatly pulled-out and bear only scattered teeth, thus modifying the general appearance to a marked degree.

Desmarestia aculeata is common on rocks at and below low tide in spring and summer and may be found readily on either side of the Cape. It is not nearly as acid as *Desmarestia viridis*.

These seaweeds have a complex and unique structure. Each species starts out as a microscopic filamentous axis bearing delicate lateral filaments or hairs along its two opposite sides. Secondary development of cells along the axis then encases it and the lateral hairs with an increasing thickness of small-celled tissue. In some species, such as *D. viridis*, the resulting plant axis is round. In others, as *D. aculeata*, it is obviously flattened, but not greatly expanded in width. In still others, such as those found on our West Coast, a great strap-shaped plant with strap-shaped branches results. Kingsbury once collected a specimen of one of the West Coast species that approached 15 feet in length. Its flat straps were nearly four inches wide. That such closely related plants can be so considerably different in aspect is worth emphasizing. *Desmarestia* and its relatives are common along the shores of Antarctica.

The life cycle of *Desmarestia* and its few relatives is similar to that described for the Chordariales. The *Desmarestia* you collect is a diploid plant producing haploid swarmers (zoospores) after a reductive division in relatively unspecialized unilocular structures on its surface. The other phase in the life cycle is filamentous, microscopic, and probably overwintering. Both species of *Desmarestia* on our coast are annual plants found in spring and summer only.

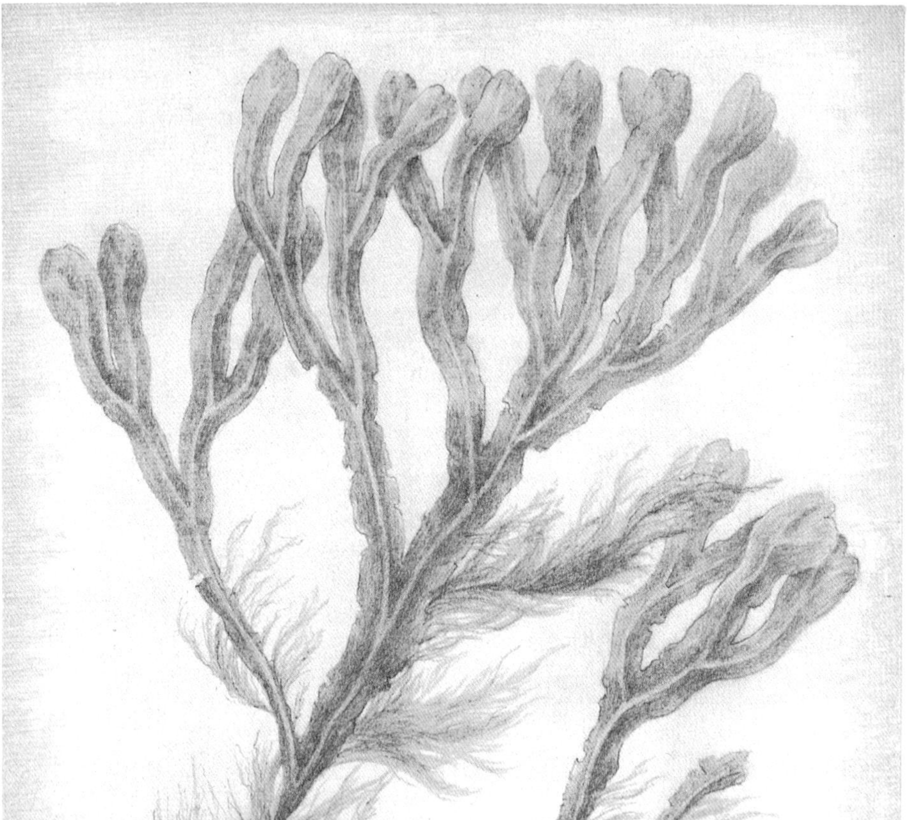

Isthmoplea sphaerophora on *Fucus* X 1.0

Isthmoplea sphaerophora

Dictyosiphonales

Derivation: "wide isthmus."
Range: Cape Cod northward.

This illustration represents something of an unusual find on the part of the Normans who collected this specimen from the Oyster River at Chatham. The species is rare at best and not expected on the south side of the Cape, since its reported distribution is farther north. *Isthmoplea* makes a tuft of what appear to be *Ectocarpus*-like filaments extending two inches or so from the substrate, usually some other alga. Under the microscope however, two differences from *Ectocarpus* are immediately apparent. Cells in the older parts of the plants develop longitudinal walls so that the plants are no longer strictly filamentous (defined as a single series of cells end to end), and secondly, the unilocular sporangia are large, stalk-less, conspicuous spheres attached laterally to the axes.

The branching pattern also is unusual, a characteristic which can be determined with a good hand lens and some care. Lateral filaments are common, diverging from the axis at nearly right angles, and usually occurring in pairs, one opposite the other.

Occasional longitudinal walls are found in the otherwise strictly filamentous *Pilayella*, especially at the points of branching. *Isthmoplea* is thus like a *Pilayella*, only more so. The remaining members of the Order Dictyosiphonales, discussed in the following pages, continue elaboration of this structural innovation (evolutionary sense), the longitudinal wall. The plant body in each case is built up of cells which have originated by divisions in all three planes of space. The assemblage of seaweeds in the Dictyosiphonales parallels the general external structure, reproduction, and life cycles of those in the Order Chordariales, but the fundamental internal structure is entirely different: cells derived from divisions in two or three planes of space in the first vs. aggregated filaments derived from cell divisions in just one plane of space in the second.

You may have trouble pronouncing the name of this seaweed. Purists in classical languages would be happy to supply appropriate rules. Pronounciation of scientific Latin, however, is not legally bound by such rules. You may pronounce these names any way you wish and no one can call you wrong (though you may have some trouble communicating).

Punctaria tenuissima on *Zostera* X 1.0

Punctaria tenuissima

Dictyosiphonales

Derivation: "dotted."
Range: North Carolina northward.

This seaweed looks much like scraps of dirty brown paper excelsior. The ribbon-like blades are usually one to a few inches long, about one quarter inch wide, and regular in width, except tapering at the base to the point of attachment which is a small button-like disc. The ribbons feel papery between the fingers and are not slippery. When young, their edges are clothed with a large number of simple filaments; these break off as the plant ages. The usual place to find this species of *Punctaria* is attached to the blades of eelgrass, often profusely as shown here.

The blade of *Punctaria tenuissima* is composed of a sheet of cells joined laterally, and is usually two cells in thickness throughout. The cells are all similar and contain equal amounts of the photosynthesic pigments. *Pilayella, Isthmoplea,* and *Punctaria* present a nice series of increasing complexity and size as the result of an increasing number of cell divisions in two and three planes of space, as compared with the basic filamentous type, such as *Ectocarpus,* from which they probably evolved. This series is carried further by the seaweeds described on the following pages.

Somehow, recognizing a pattern in a series of organisms gives the biologist a feeling of "understanding" the individual members of the series better than knowing exactly the same things about each, but without discerning any pattern in them. It is always fun, and sometimes instructive, to propose a hypothetical organism as the next logical step in a series with pattern. What characteristics would you propose for a seaweed of greater structural complexity but related to *Punctaria tenuissima?* How big can something like this get? What structural problems are presented by increasing size? What innovations are logical solutions to such structural problems? Are any structural problems associated with increasing size potentially insoluble?

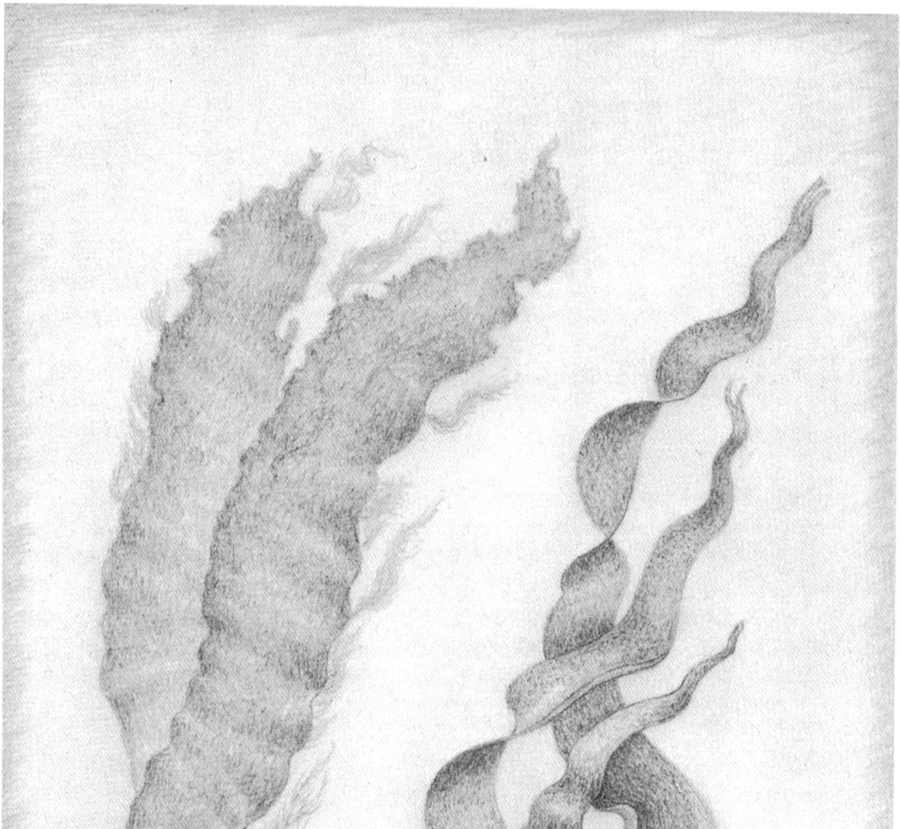

Punctaria plantaginea X 0.50 *Punctaria latifolia* X 0.50

Punctaria plantaginea Dictyosiphonales

Derivation: "dotted."
Range: Virginia northward.

This is the largest of the three local species of this genus. It grows on rocks and large seaweeds and is very common in spring and early summer. The blade is tough and persists through the summer, especially in deeper water, but becomes tattered and covered with epiphytes as shown here, and you are unlikely to find a "clean" specimen after early summer.

The blades range from about six inches to a foot in length and are usually about two inches wide, with mostly parallel, unwavy edges. They taper to a very short stalk at the point of attachment which is a small, button-like disc. The outer end of the blade is blunt or torn into segments, becoming more shredded as the blade ages. Mature blades have something of the consistency, appearance, and "feel" of the tongue of a shoe. They become very dark brown or nearly black with age.

To get a good look at the structure of this seaweed, it is necessary to cut a thin strip from it (cross-section) that can be made to lie out on a glass slide so that you can look down on the cut surface and see the internal arrangement of cells across the thickness of the blade under the microscope. The blade is composed of cells dividing from one another to a thickness at maturity of four to seven cells. These are all nearly alike in structure and contents (there is no clear differentiation into cortex and medulla). Each surface of the blade bears scattered tufts of filaments and groups of reproductive organs. These can be seen with a hand lens in a young plant before it becomes loaded with epiphytes. The dotted appearance that these tufts give to the blade accounts for the seaweed's name.

The species of *Punctaria* are annual seaweeds, disappearing in winter. *Punctaria latifolia* of nearly the same range as the other two, grows commonly on eelgrass and coarse seaweeds. It is lighter in color, more delicate in substance, and somewhat smaller in size. A feature useful in distinguishing it is the fact that the base of the blade usually tapers very abruptly but symmetrically to the stalk as compared with the more gradual taper in *Punctaria plantaginea* and *Petalonia fascia* (page 93). The blade of *Punctaria latifolia* occasionally has wavy edges. Secure identification of less than typical specimens of all three of these often requires studying a section of the blade under the microscope.

Asperococcus fistulosus X 0.75

Asperococcus fistulosus

Dictyosiphonales

Derivation: "rough berried."
Range: North Carolina northward.

This unbranched, tubular, inflated seaweed is another of the group that makes its appearance in late winter, fruits in early spring, and disappears entirely as the summer progresses. It grows, usually many individuals together, on rocks, wood, and particularly on *Fucus* and its relatives at and below the level of low tide, and is favored by waters of less than full exposure to surf. Individual plants are from a few inches to more than a foot in length, and about a half inch in diameter. Except for the absence of constrictions, plants of *Asperococcus* are not unlike those of *Scytosiphon*. The basic construction also is similar. A cortex of small, densely pigmented cells is external to a medulla of larger, nearly colorless cells. The medulla is solid in very young plants and at the stalk-like base of other plants but becomes torn apart in the swollen portion as the plant matures.

A characteristic that is useful in identifying *Asperococcus* in the field (we have only the one species) is the nature of its surface. The plant is almost always fruiting. The fruiting structures are gathered in clusters on the surface and are interspersed with short filaments that grow outward from the surface. These clusters may be seen with hand lens and give a characteristic "feel" to the plant. When you run it lightly through your fingers it feels like very fine sandpaper. While *Scytosiphon* usually is not distinctly slippery, neither does it have quite the feel of sandpaper. The clusters of fruiting structures are even more markedly developed in older specimens, forming conspicuous bumps, whence the name of the alga.

It may be chance that several members of the Orders Dictyosiphonales and Scytosiphonales, for example *Punctaria, Asperococcus, Petalonia,* and *Scytosiphon,* are winter and spring seaweeds, but this seems a good place to call attention to the phenomenon of seasonal succession. When you first start visiting the rocky seashore, the variety of conspicuous seaweeds at each place will impress you. Only after you have both learned the common seaweeds well, and have visited the same locality repeatedly over a period of at least several weeks, can you expect to see effects of seasonal change. Many seaweeds, especially the larger browns and reds, are perennial, and a population replaces itself slowly and steadily, showing little obvious change in the absence of some unexpected catastrophe. It is among the shorter-lived annuals that the seasons are measured. The early-season brown seaweeds named above tend to be replaced by short-lived red seaweeds such as *Ceramium, Polysiphonia,* and *Nemalion,* which appear in early summer and increase in abundance as the season wears on.

Dictyosiphon foeniculaceus X 0.50

Dictyosiphon foeniculaceus

Dictyosiphonales

Derivation: "netted tube."
Range: Virginia northward.

Dictyosiphon is a large branched and rebranched bushy plant, with firm, narrow axes, not more than an eighth of an inch in diameter even in the oldest parts, which may become slightly or irregularly hollow with age. The color is usually a good, obvious brown. *Dictyosiphon* differs from *Desmarestia viridis*, with which it might be confused, by being stiffer. Its branching pattern is random, but the main axis is usually distinct through most or all the height of the plant. It is not conspicuously overtopped by the major branches as in *Chordaria* and *Acrothrix*. This characteristic is well shown in the illustration. The branches` have smooth, but not slippery surfaces, and taper toward the tip where they may be almost threadlike. Young branches may bear a fur of fine, deciduous filaments.

Dictyosiphon grows on various of the larger seaweeds, especially *Chordaria*. You can usually tell easily where the *Chordaria* ends and the *Dictyosiphon* begins under such circumstances by the difference in color, *Dictyosiphon* being much lighter brown. It lives year-round, and is particularly common on the southern shores of Cape Cod.

Dictyosiphon is built of cells dividing from one another in three planes of space, thus forming a solid tissue. The outer cells become small and densely colored, constituting a cortex, while the inner cells are larger, colorless, and collectively constitute a medulla. The specialty of structure that characterizes *Dictyosiphon* is that its main increase in length is brought about by the divisions of four distinctive cells at the tip of each axis. In other members of the Dictyosiphonales, growth is mostly random, or localized in less specialized cells.

Dictyosiphon and most other members of the Dictyosiphonales are diploid plants as you collect them. They bear unilocular sporangia in which reduction divisions take place and from which are released haploid swarmers (zoospores). In the most fundamental situation, these give rise to microscopic filamentous haploid plants which bear swarmers that fuse (gametes), and the progeny of which sooner or later re-establish the diploid stage again (see diagram, p. 71). In the alternation of a large obvious diploid phase, with a microscopic haploid phase, the Dictyosiphonales are similar to the Chordariales. They differ in the basic characteristics of their structure—interwoven filaments in members of the Chordariales, and solid, three-dimensional cellular tissue in the Dictyosiphonales.

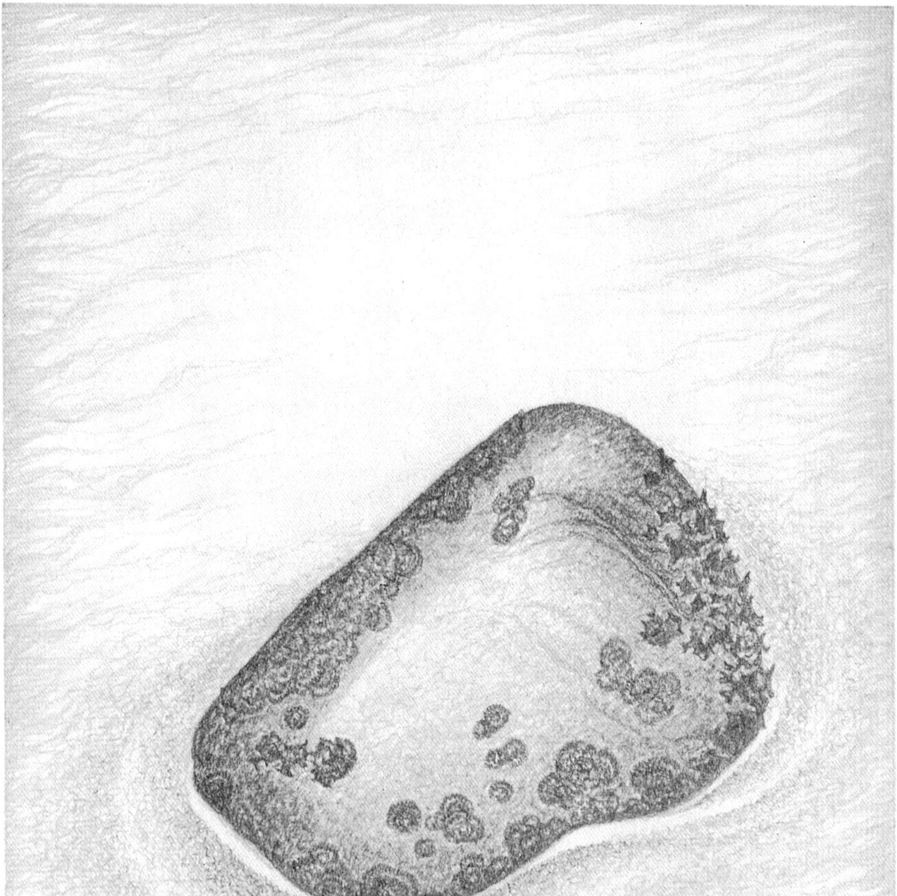

Ralfsia verrucosa X 1.0

Ralfsia verrucosa Scytosiphonales

Derivation: honors John Ralfs, an English botanist.
Range: Virginia northward.

This common species of *Ralfsia* grows on rocks or woodwork between tides or just below low tide. Individual plants look like heavy dark brown or black stains or patches of tar on the rock. In fact some *Ralfsias* are almost indistinguishable from a tar spot on casual inspection. Kingsbury has seen an eminent phycologist (a botanist specializing in algae) make this mistake. The difference is immediately obvious when you try to scrape off a tar spot; it gums up your knife. Submerged in tidepools, *Ralfsia verrucosa* often resembles a brown crustose lichen.

A young *Ralfsia* starts its growth with a radiating layer of filaments prostrate against the substrate. As it develops, every cell of the prostrate filaments supports an erect, unbranched filament that grows from it. Erect filaments are produced so densely that they lie tightly against each other and the *Ralfsia* plant becomes a crust as many cells thick as the erect filaments are high. *Ralfsia verrucosa.* gets quite thick. Plants are perennial, grow slowly, often displaying concentric zones of growth, the oldest at the center, and they gradually thicken with age. Young plants are usually circular while older crusts may become very irregular in outline. A very old plant becomes crumbly like a piece of bark. It may bend irregularly and lift itself free of the substrate in places. Such fragments are easy to collect if you are lucky to find them.

Ralfsia has a life cycle and reproductive pattern similar to that of *Ectocarpus* and *Pilayella.* Any given plant may be haploid or diploid. To tell, one must see what type of reproductive structure, unilocular or plurilocular, is produced on it.

A complication when trying to recognize *Ralfsia* is that other brown algae such as *Petalonia* and *Scytosiphon* have stages closely resembling *Ralfsia* in their life cycles.

Petalonia fascia X 0.75

Petalonia fascia

Scytosiphonales

Derivation: "leaf-like."
Range: entire East Coast.

A fundamental characteristic distinguishing *Petalonia* from *Punctaria* which resembles it, has to do with its cellular construction. The blade in *Petalonia* is about as thick as it is in *Punctaria* and is composed of cells that have divided from one another, just as in the latter, but they are segregated into two areas composed of cells of distinctly different size and color. The cells of the surfaces are small and contain pigment-bearing chloroplasts, hence are dark brown in color. The cells of the center, sandwiched between, are much larger, and mostly lack chloroplasts, hence are nearly colorless. As in other seaweeds in which a similar separation of cell type and color occurs, the outer, colored zone is termed the cortex and the central nearly colorless layer, the medulla. Possession of "tissues" of this sort is considered an advanced evolutionary characteristic. *Punctaria* lacks clearly defined differentiation into tissues.

Petalonia, contrary to most seaweeds, grows best in winter, maturing and disappearing in early summer. The ribbon-like plants usually occur in clumps, attached to rocks or wood, high in the intertidal zone. Each blade is olive or "middle" brown, smooth, leathery, and tapered at both ends. Plants range from a few inches to nearly a foot in length. The lower end of the blade narrows into a cylindrical stalk which in turn terminates in a button-like disc at the point of attachment to the substrate. In some varieties, the blade is markedly wider than in others and more easily confused with that of *Punctaria.* Note that under such circumstances the narrowing at the base of the blade tends to be markedly asymmetrical in *Petalonia,* in contrast to *Punctaria* in which it is quite symmetrical. Both *Punctaria* and *Petalonia* can easily be distinguished from young *Laminaria,* because even in very young plants of the latter, the attaching structure or holdfast at the base of the stalk is divided into obvious branches.

In higher plants and in animals, tissues are of rather precise composition. Applied to algae, however, the term "tissue" is used much more loosely. Thus we find tissues of cortex and medulla in both Chordariales and Scytosiphonales, though in the former they are composed of interwoven filaments, while in the latter of solid cellular mass.

Scytosiphon lomentaria X 0.75

Scytosiphon lomentaria Scytosiphonales

Derivation: "whip-like tube."
Range: entire East Coast.

A well developed *Scytosiphon* looks like a chain of sausages. Each plant is an unbranched, inflated tube, constricted at more or less regular intervals. *Scytosiphon* forms luxuriant growths attached to rocks at or below low tide. It is an annual plant which has its best development in the winter, persisting through the summer in some places but disappearing entirely in most Cape Cod locations as the water warms. Sometimes plants grow so closely together that a clump looks like a single plant branching from the base. Each individual may be from a few inches to a foot and a half long and most are about half an inch wide. You can easily determine that the swollen portions are hollow by squeezing them.

Scytosiphon starts life as a solid cylinder of cells. It increases in size by repeated division of these cells and their progeny in three dimensions of space. As in *Petalonia,* the cells toward the outside become smaller and densely colored, compared with the large, almost colorless cells of the central area, and thus cortex and medulla are differentiated. The cells of the cortex continue to divide, forming stresses around the circumference of the plant which sooner or later pull the interior apart. This is what accounts for the hollow center. The "knots" along the chain of sausages are simply places where the center has not pulled apart but remains solid.

A well developed *Scytosiphon* is not likely to be confused with any other seaweed. Superficially it may resemble an *Enteromorpha intestinalis,* but the wall is thicker, the constrictions more precise, and the color—deep brown compared with bright green—is entirely different. Young *Scytosiphon* in which the inflated character is not yet well developed, or older plants which for some reason do not show good constrictions, however, may be confused with young *Chorda* or *Asperococcus.*

Scytosiphon is widely distributed throughout the world; finding its familiar and distinctive form among an assemblage of strangers on some distant shore can be like finding an old friend.

Both *Petalonia* and *Scytosiphon* form their upright phases during the winter and "disappear" by assuming a prostrate, *Ralfsia*-like phase in summer. This seasonality is controlled by changes in temperature, changes in the length of the daylight period, and abundance of herbivores.

Chorda filum X 0.10

Chorda filum

Laminariales

Derivation: "cord-like."
Range: New Jersey northward.

In late October, off the shores of Scraggy Neck in Buzzards Bay (delightful names) the large number of venerable old plants of *Chorda filum* still around when most of the summer forms are gone is impressive. Even the red epiphytes which often crowd the surfaces of *Chorda* as the summer progresses disappear, leaving only an occasional almost yellow-rusty *Ectocarpus* on the *Chorda*. The *Chorda*, being annual, will soon also disappear, but perhaps its coarse, firm, thick structure allows it to last longer than the more delicate forms. The water is getting cold, and the reds and greens of the summer seaweeds have disappeared leaving the perennial and winter-growing brown seaweeds to dominate the color of the water, almost in reflection of the change from summer greens to fall reds, and finally to the gone-by brown colors of autumn on land.

Fully developed individuals of *Chorda* are easily recognized. Each is attached to an appropriate substrate usually a stone or a shell, by a small, discoid pad or holdfast. The unbranching cylindrical tube which arises from this is thin and solid at the bottom, gradually expanding and becoming hollow upwards, to about a half inch in diameter, and if the plant is not broken, again tapering toward the upper extremity. The majority of plants shown here have been broken; three have not. A small plant might be confused with an unconstricted *Scytosiphon* or an *Asperococcus*, but the substance of the plant is thicker and firmer in *Chorda* than in either of the other two. The cavity is relatively small in proportion to the tissues surrounding it, and is traversed by diaphragms of colorless tissue at intervals but is not constricted at these points as in *Scytosiphon*. The surface is smooth and regular. It is clothed, in young plants, with a fine fur of long colorless hairs or filaments. These give the plant an indistinct outline as it grows in the water and seem to disappear when it is taken out because they come to lie invisibly against the surface. They are shed entirely as the season progresses.

Chorda lives below low tide down to considerable depth. The hollow tube is filled with air or respiratory gases and extends more or less upright in the water. Some plants may be ten feet or more in length, though the majority are half that or less. The tips of the plants are usually below water but in shallow areas may come to lie out on the surface as the tide goes down.

Chorda and the following brown algae through *Alaria* belong to the Order Laminariales, the kelps.

Chorda tomentosa X 0.25

Chorda tomentosa

Laminariales

Derivation: "cord-like."
Range: Long Island northward.

In contrast with the more common *Chorda filum* described on the preceding page, *Chorda tomentosa* is a winter plant of cold waters. Classes from Woods Hole have repeatedly collected it in June at the Cape Cod Canal jetty at Sandwich and this is also where the Normans found the specimen illustrated here. Even there it is not common and must be sought carefully. It is attached at some depth below low tide to a rock or other solid substrate. In general appearance, it is like *Chorda filum*. In fact, out of water they are difficult to tell apart. Under water, however, the major difference is immediately obvious. The surface of *Chorda tomentosa* is densely covered with long, very dark brown, almost black hairs. Examination of the plant by microscope shows that these are unbranched filaments growing out of the cells of the surface. Every cell of these hairs contains chloroplasts and is densely provided with brown photosynthetic pigments in *Chorda tomentosa*. The cells of the deciduous filaments of *Chorda filum*, on the other hand, never have this appearance. When *Chorda tomentosa* is removed from water, the surface hairs bend at the base and lie flat in the meniscus of water closely along the sides of the tube. Thus they seem to disappear.

The colored surface "fur" of *Chorda tomentosa* is present throughout the life of this species which in our waters appears in winter, fruits in spring, and disappears in early summer. The deep brown color of the "fur" indicates that its function is to aid in the photosynthesis of the plant. The superficial filaments of *Chorda filum*, on the other hand, secrete slime on the surface of the plant. Of what advantage is it to the plant to use its resources in forming colorless hairs and in continuously secreting slime on its surface? Slime production, a very common phenomenon among seaweeds, helps to keep the surfaces of the plant free from epiphytes. Attachment is difficult on a surface which continuously dissolves. The intense competition for "sitting-down-space" at the edges of the ocean and the resulting phenomenon of epiphytism has already been discussed (page 65; also 207). Plants which become heavily epiphytised are shaded by the plants growing on them, and their ability to accomplish photosynthesis is reduced proportionately. A heavy burden of epiphytes also makes them more susceptible to being torn away by the suck and surge of waves and storms.

Laminaria saccharina X 0.10

Laminaria saccharina

Laminariales

Derivation: "a thin blade."
Range: New Jersey northward.

Laminarias are so conspicuous and distinctive that they have won a common name—"kelp". They are the most obvious seaweeds wherever they occur on Cape Cod. *Laminaria saccharina* is quite common in the warm waters of the southern shore of Cape Cod and the Islands, forming populations attached to pilings, stones or especially to large mussels *(Mytilus)* as shown here. All species of *Laminaria* are plants of the subtidal zone where they make the upper part of that zone conspicuously different from the zones above it. *Laminaria saccharina* grows typically at depths between two and fifty feet. You can best appreciate these plants by swimming among them with a face mask.

Laminaria saccharina looks like an over-developed razor strop. Large plants may be ten feet long and six to ten inches broad. The blade is frayed and worn at the outer end where it may have a sparse covering of epiphytes. At the other end it narrows slowly or abruptly into a cylindrical stalk which in turn terminates in the holdfast. The length of the stalk is roughly related to the depth of the water in which the plant is growing. It may be a few inches or several feet long. The plant is perennial. In winter the blade is quite thick. The blade produced in spring and summer is thinner and has ruffled edges as shown in this illustration. In late spring you may find plants with an old, unruffled blade at the tip separated from the new summer blade being produced below.

The holdfast of a *Laminaria* plant is similar to the holdfasts of most members of the Order Laminariales. An inverted basketlike cluster of branches (called haptera) grows downward from the bottom of the stalk over the rock or shell to which the plant initially attaches. Germlings are relatively indiscriminate in their choice of substrate, attaching to rocks, shells and other hard surfaces of various sizes. Many attach to mussel shells; the mussel in turn is held to a rock by means of the tough byssus threads which it secretes. As the kelp grows large it may become more than the mussel can hold. In a storm the whole mass is pulled free from the rock, and both the mussel and the kelp eventually die. Rarely do you find a holdfast which has been pulled free from a mussel, or a stalk which has broken. The attachment of the mussel to the substrate is usually the weakest link in this association.

In the Order Laminariales, as in the Chordariales and Dictyosiphonales, the large plant as collected is the diploid phase in the life cycle, producing swarmers in unilocular structures that grow from the surface of the mature blade. Haploid in chromosome number, they germinate into the alternating phase consisting of microscopic, few-celled, branching filamentous plantlets. These are the sexual plants and they can be separated visually into either male or female. Male plants are composed of more and smaller cells than the female and branch more copiously. Both sexes produce single-celled reproductive structures, each of which releases a single gamete. The male gametes are small and motile. The female gametes are large and non-motile. Where thus differentiated, as in *Laminaria*, the male gametes are commonly called sperm and the female gametes, eggs.

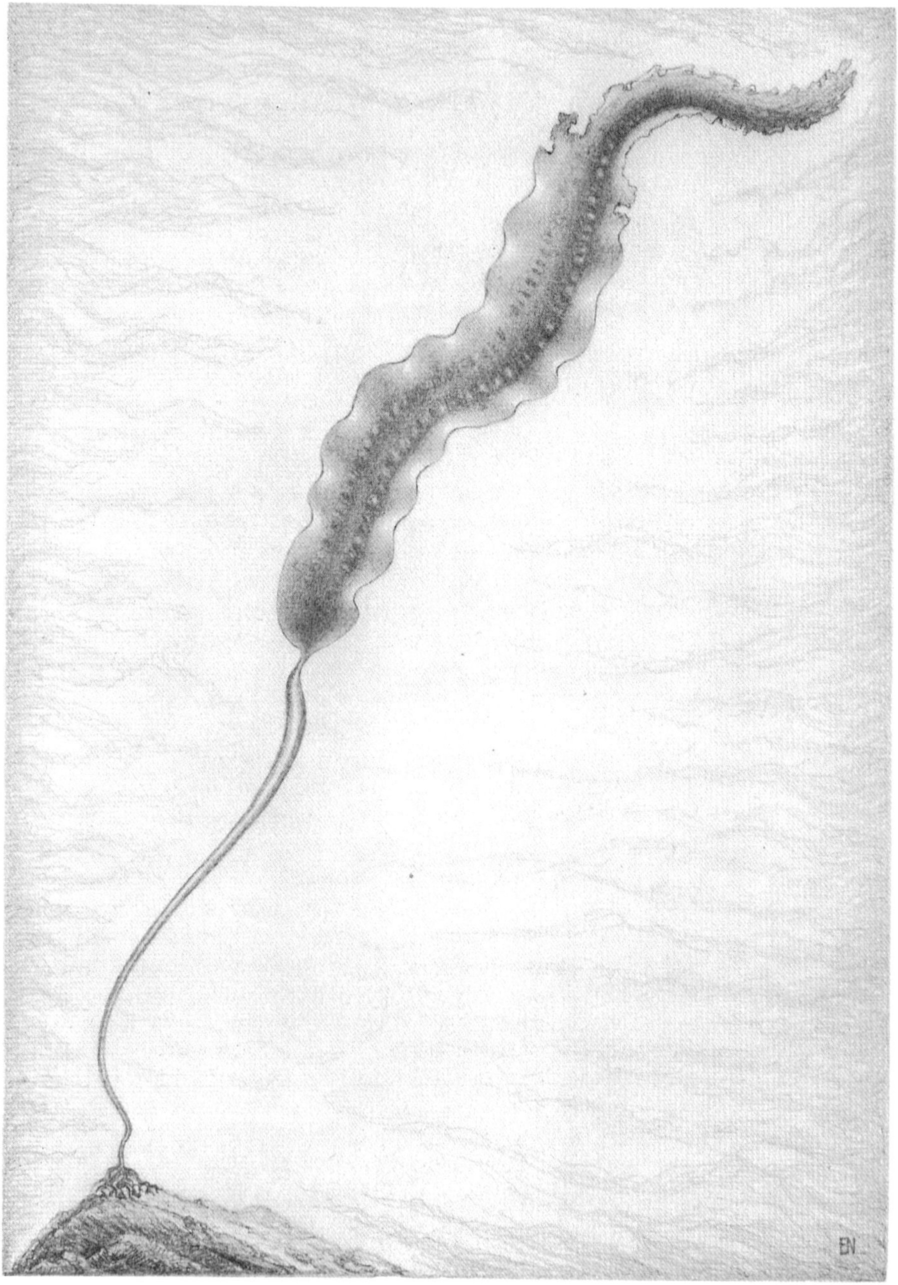

Laminaria longicruris X 0.10

Laminaria longicruris — Laminariales

Derivation: "a thin blade."
Range: Long Island northward.

This kelp is rare at Cape Cod; more common northward. If you find it at all, it will probably be cast up on a beach, particularly of the north or east side of the Cape, after a storm. The Normans found this one at Nauset Inlet, Orleans. It looks like the far more common *Laminaria saccharina*, but typically is larger. The greatest difference is in the structure of the stalk, which is hollow, gradually enlarging and becoming distinctly inflated from the holdfast upward to the growing region between stalk and blade.

A blade of *Laminaria* resembles a slowly moving conveyor belt as new tissue is added at the growing region between stalk and blade and old tissue is worn away at the outer end. This pattern of growth provides an easy means of estimating growth by punching holes near the base of the blade and measuring how far the holes have moved from their original location at a later time. This method was applied by K. H. Mann to study the seasonal pattern of kelp growth (*Laminaria longicruris, L. digitata, Agarum cribrosum*) in Nova Scotia. He estimated that one plant completely replaces its blade several times each year, and surprisingly, the greatest growth occurs in late winter when water temperatures reach their lowest for the year. The kelps take advantage of the high concentrations of dissolved nitrogen compounds in the seawater in winter, which they require for the synthesis of new tissues. As the water warms in the spring, these critical nitrogen sources are consumed by planktonic algae during their "spring bloom", and the supply remains depleted until the next winter. Thus despite the warmer temperatures and greater sunlight for photosynthesis during the summer, kelp growth is limited by the lack of a critical nutrient. At the same time, kelps are able to make and store energy-containing organic material (carbohydrates, not requiring nitrogen) during the summer when conditions are more favorable for photosynthesis and then use these food reserves for their late winter growth when light availability is least.

These important seaweeds have adapted to conditions in cold, northern oceans that seem at first to be far from favorable for plant growth. In fact as other kelp beds have been studied, the conclusion has emerged that they are among the most productive biological systems on earth based on how much plant material is produced on a square yard of ocean floor. In this respect, their annual yield is similar to intensively managed agricultural crops and tropical rainforests. Kelp beds, however, are limited in their geographic extent, requiring shallow waters of cold, rocky shores.

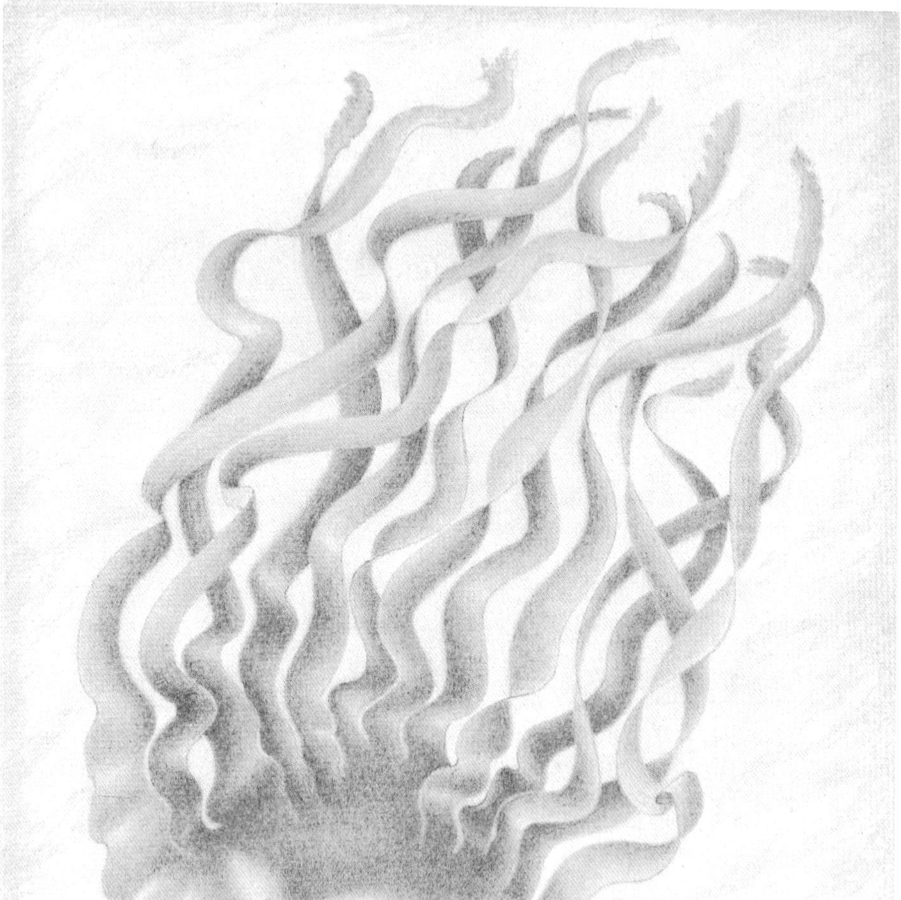

Lamninaria digitata X 0.45

Laminaria digitata

Laminariales

Derivation: "a thin blade."
Range: Long Island northward.

Laminaria digitata is less commonly found than *Laminaria saccharina*. It occurs in more limited areas and in somewhat colder waters. On Cape Cod, look for it in the cold, stony areas such as off Gay Head or the Sandwich canal jetty. *Laminaria digitata* is basically like other kelps, except that the young blade becomes much wider, and then splits along its length several times so that the fully developed blade is divided into a half dozen or more strap-like pieces. The stalk in this species tends to be quite stiff, almost woody.

Kelps possess the most complex and varied cell types of any algae. Their stalks and blades are colored in the surface layers, colorless and solid within, but the whole is rather more complex than the cortex and medulla of other seaweeds, and beyond easy description here. The way the plants grow in length is unique among all plants. The growing zone, that is, the place where cell division is most active, is located just where the stalk joins the blade. If you cut through a plant at this spot and compare what you see under a microscope with a similar section through a mature part of the stalk, you will find little difference in cell types. Close observation discloses that actual divisions of cells in the growing region are limited to the cells on the very outside; the ones inside are merely being stretched. How does a plant grow in length, and not in width, when divisions are thus localized? Visualize a soft glass tube heated and pulled out in a flame. The tube increases in length at the expense of becoming thinner at the heated spot. Thus it is with kelps, except that new cells are continually formed at the point of elongation so that the stalk maintains approximately its same diameter throughout. The great growth on the surface in this zone causes the internal cells to be stretched increasingly until they look like fine, tubular threads at the center of the stalk. The structure of the stalk and blade is basically similar. The blade undergoes differential expansion to become flat instead of cylindrical. As the growing region adds to it at the bottom, it continually wears away at the tip. The splits in the blade of *Laminaria digitata* are not random, but are determined by lines of weakness inherent in the structure of the blade as it develops (see page 107).

The original rich native populations of shellfish and fish have been depleted or extinguished by intense harvesting in many marine areas. What about seaweeds? We once calculated that a single kelp plant can produce over 500 million unilocuilar structures on its surface and release over a trillion swarmers into the surrounding water during a single reproductive season. These numbers are multiplied again by the alternating sexual stage in the life cycle. Potentially an astronomical number of new *Laminaria* individuals can be derived from a single diploid *Laminaria* plant. Since over its lifetime each *Laminaria* is successful in reproducing itself only once (populations remain static), think of the fate of the vast majority of reproductive cells that never develop into adults but die or are eaten by filter-feeding animals. Seaweed populations are endangered far more by human pollution than by human beings collecting responsibly.

Agarum cribrosum X 0.25

Agarum cribrosum

Laminariales

Derivation: from an old word for mushroom.
Range: Cape Cod northward.

Several of the kelps are called "devil's aprons," but this particular seaweed perhaps best deserves the name because of the holes which are its most distinctive feature. These are supposed to represent burn-spots from sparks, an occupational hazard in the Devil's abode. Other names sometimes applied to *Agarum* are "shotgun kelp" and "sea colander." Applying common names to seaweeds is risky, however, because of the geographic variation and lack of consistency in their use.

Agarum can be distinguished from all other kelps by the holes alone. These are round to oval, and regular in distribution and size. They are well shown in the illustration. In addition to this distinguishing characteristic, the plant possesses a typical branched holdfast, an indistinctly thickened midrib region (lacking holes), and a moderate stalk. It grows below low water. On Cape Cod, which represents the southern limit of its distribution, plants are rare, and are usually found only after a northeast storm washes them onto the northern beaches. Plants are usually torn or severely abraded at the outer edge. The more resistant midrib area wears away more slowly than the rest. Most specimens are a very dark brown, almost black in color. Sometimes, as for an old *Fucus* plant, the color may have overtones of red in it, similar to the color of an autumn oak leaf.

Agarum is a seaweed of the Arctic extending down the coasts of North America to Cape Cod in the east and Washington State in the west. It prefers deeper waters than most other kelps. As attached organisms, most seaweeds would seem like sitting prey for herbivores. *Agarum* illustrates well the defenses employed by many of the larger brown algae against the pressures of such herbivores as sea urchins or snails. The first line of defense is physical toughness against the penetration and subsequent damage by animal feeding structures. The second line of defense is chemical. Most brown algae, and *Agarum* in particular, produce complex compounds (polyphenols) that deter grazing. This combination of defenses is very similar to the defenses of tropical rainforest plants against insects and other animals.

The holes in an *Agarum* blade, like the splits in a blade of *Laminaria digitata* are not random events. Each is predetermined in the development of the plant by overgrowth of the outer or cortical tissues in a particular way which results in a point or line of weakness in the developing blade. Wave action causes the blade to break through at such places, and further growth and elongation of the blade causes the initially very tiny holes in an *Agarum* to become increasingly larger until they attain the size characteristic for the mature plant. You can see this process taking place by comparing the small holes near the area of active growth (between stalk and blade) with holes in the older parts of the blade farther out.

Alaria esculenta X 0.25

Alaria esculenta

Laminariales

Derivation: "winged."
Range: northern Cape Cod northward.

This seaweed is easy to recognize and you may consider yourself lucky if you find it on Cape Cod or the Islands though it is quite common to the north. It is the only kelp (of ours) to possess a distinct, sharply edged midrib. Furthermore, except for a short time each year, it bears characteristic secondary reproductive bladelets, and in this is like no other of the East Coast kelps. The mature plant, up to ten feet long and about a half foot wide, has a branched holdfast, cylindrical stalk, and expanded blade like a *Laminaria*. The blade, however, has an obvious, thick midrib, about half an inch or a little less in width with exactly parallel edges. The midrib is much thicker than the rest of the blade and persists longest as the vegetative tissues wear away at the outer extremity. The wavy, thin margins of the blade tend to tear easily, nearly perpendicularly to the midrib in the outer part of the blade which usually looks quite tattered and worn.

The stalk is thin and stiffish. It lasts several seasons and produces a yearly crop of small, thick, lateral blades, somewhat irregular in outline, in two rows, one down each side. These constitute the site of swarmer production. The main blade is entirely sterile. The substance of each reproductive blade is largely used up in swarmer production, and they are shed after the swarmers have been released, leaving bumps on the stalk where they were attached. These structural details are well depicted in this illustration. Separation of reproductive and vegetative functions into specialized parts of the plant is considered an advanced characteristic. In some of the west coast kelps, the reproductive blades have evolved to the point where they are larger than the primary, vegetative blade. In attaining this end, evolution has proceeded nearly full circle; the final result is a plant very similar to that which represents the starting point. These west coast forms look not unlike a large *Laminaria digitata* (in which reproduction is scattered throughout).

In each case, the species name of *Laminaria saccharina* and *Alaria esculenta* has something to do with edibility. In Ireland and Scotland, the midribs and reproductive blades of *Alaria esculenta* are harvested and sold in stores to be eaten. Similarly, kelp blades when dried often form a copious white powder on the surface. This is a carbohydrate material which is mildly sweet to the taste. Appearance of the carbohydrate powder is not limited to *L. saccharina*, but may also occur with a variety of other kelps.

In the Orient *Laminaria* (sold in stores as *kombu* in Japan and *haidai* in China), and another kelp *Undaria* (*wakame* in Japan), are cultivated and commercially harvested as foods and used as dietary supplements to supply iodine. In North America, the principal use of kelps is as a source of the gum, algin (see page 170), which comes primarily from the harvest of natural populations of the giant kelp, *Macrocystis*, on our west coast.

Fucus vesiculosus X 0.75

Fucus vesiculosus

Fucales

Derivation: from old Greek, meaning "seaweed."
Range: North Carolina northward.

This is probably the most common of our intertidal seaweeds. *Fucus* and its almost equally common relative *Ascophyllum,* are generally called "rockweeds". If a plant has discrete, mostly paired air bladders and a distinct midrib, as shown in the illustration, it is certainly *Fucus vesiculosus.* If it lacks air bladders, it still may be this species, or it may be one of two others described on the following pages.

Fucus vesiculosus grows attached to rock, wood, or shells, in the mid to lower intertidal area, but not below low tide. A perfect, uninjured plant shows a very regular pattern of branching. Two equal branches are produced at the first bifurcation of the original axis in the young plant. Each of these soon bifurcates again at approximately the same time, so that the two original branches separate into four of equal size and position. The process is repeated several more times and the adult plant is thus potentially a regular and planar collection of bifurcating axes ("Y"s). Injury or mischance usually prevents the development of the full pattern in any given individual. The perfect potential pattern of branching in *Fucus* is perhaps the best display of a type of branching considered on fossil evidence to be a basic type in the evolution of higher plant groups.

The youngest part of a *Fucus* plant is at the branch tips. The cell divisions responsible for growth in length of the branches represent the activities of a single cell at that location. To locate it, you must have tips of growing, vegetative branches. These are usually somewhat yellower in color than the rest of the axis and flat (in comparison with the swollen reproductive tips). Each growing tip, as seen with hand lens (or just barely by naked eye) has a flattened indentation or dimple at the outer end. Only with a microscope, however, can one see the actual dividing cell which appears as a single, large, dense cell embedded at the exact center bottom of this funnel-shaped depression. The dividing cell gives rise to three dimensional tissue which, through displacement stretching, swelling and some further divisions, differentiates into the midrib and wings of the adult structure.

Fucus plants are attached by an unbranched holdfast. Above that, except in young plant the wings have usually worn away leaving the midrib as a stalk. Air bladders, if produced, usually occur in pairs. When the plant is reproductive, which is more than half the year (winter), almost all branches terminate in swollen, mucilage-containing reproductive structures, the receptacles. The surfaces of both the vegetative and reproductive parts of the plant are covered with tiny dots which represent the mouths of walled cavities into the looser tissues beneath, and in which the reproductive organs are formed in the receptacles.

Fucus, Ascophyllum, and *Sargassum* belong to the Order Fucales and are often called "rockweeds," especially the first two.

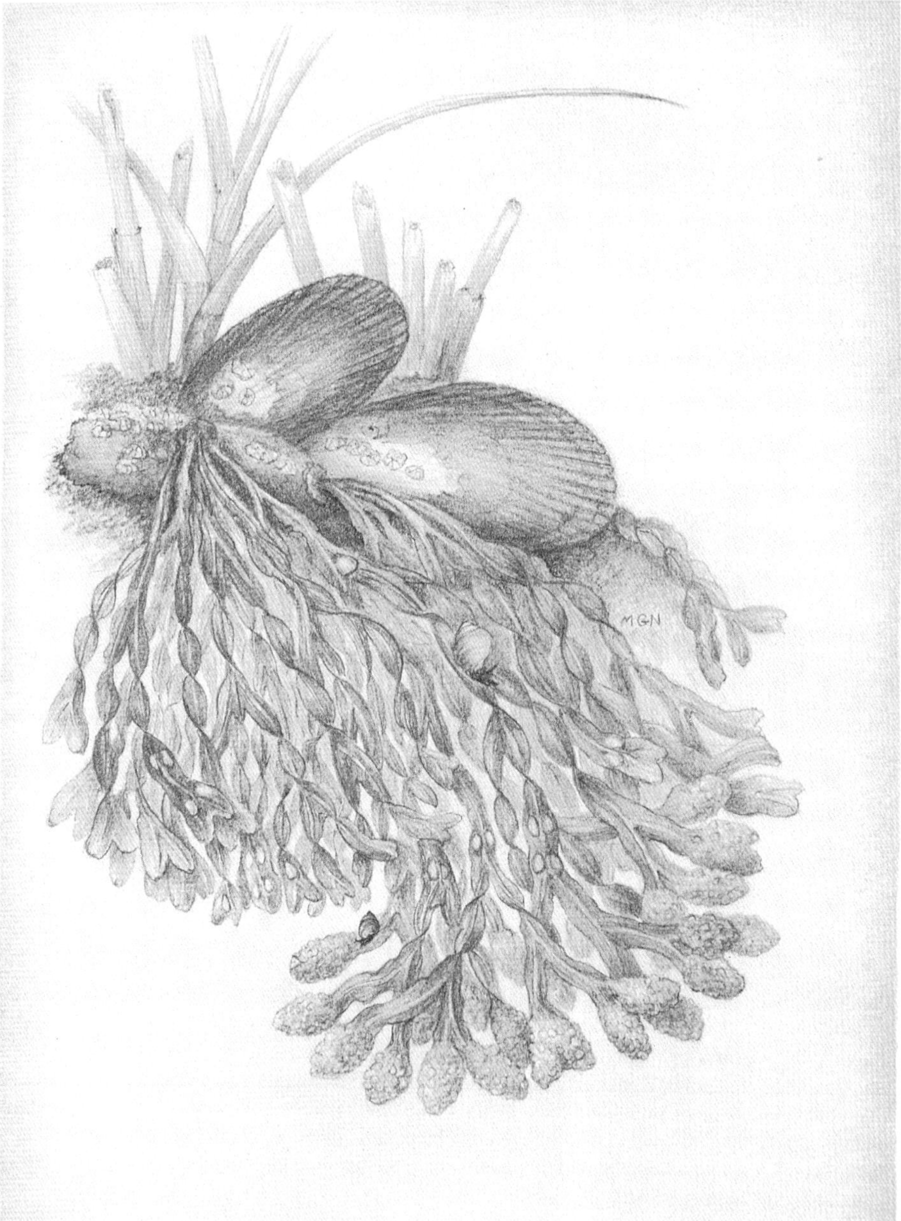

Fucus vesiculosus var. spiralis X 0.50

Fucus vesiculosus* var. *spiralis Fucales

Derivation: from old Greek, meaning "seaweed."
Range: North Carolina northward.

Note the regular and conspicuous spiraling of the branches. This specimen does not look much like the typical *Fucus vesiculosus,* but if you look more closely, you will find paired air bladders on at least some of the plants. Air bladders are found only on *Fucus vesiculosus.* This demonstrates that first impressions may be deceiving and conclusions should be made only after a careful examination.

Fucus vesiculosus grows over a wide range of conditions. It is unusual among seaweeds in being able to persist and grow under unexpected circumstances. For example, it will grow without a firm substrate for attachment, or buried in mud with only the branch tips exposed, or in estuaries with waters of reduced or varying salinities. Under such conditions the plants may assume a growth form quite different from the typical, and usually do not reproduce sexually. Often a special character of growth, here spiraling, is constant under a given set of environmental circumstances. Botanists employ several categories subsidiary to species to classify such variants. The plant illustrated here was collected by the Normans from a tidal embayment with almost marsh-like edges at Nauset Inlet, Orleans. The typical spiral variant, exemplified by this specimen, is commonly found on the banks of ditches in tidal marshes or in the marshes themselves, usually not attached to ribbed mussels as shown here, but entangled around the bases of the marsh grasses. This particular form of *Fucus vesiculosus* is considered a variety (abbreviated as "var.") The word "variety" used in this way carries a precise scientific meaning not exactly equivalent to its English usage.

Another characteristic found to marked degree in *Fucus* and also in *Ascophyllum* is the ability to regenerate branches when originals have been injured or torn off. Examine the specimens you collect for the presence of small "tabs" of tissue arising from the stalk or midrib, especially in lower parts of the plant. These often occur in conspicuous clusters. If these unorganized blades are young enough, the scar of the original injury is usually apparent. As they grow they gradually assume mature character and an apical cell appears at the growing tip.

Our open coasts where seaweeds grow experience a great deal of physical disturbance, especially during winter when rocks may be rolled about by storm waves and ice scrapes against the shore. Seaweed communities living under these exposed conditions must be able to recover from such disturbances if they are to maintain a stable population. Some years ago, Sze started a study of this process at the Shoals Marine Laboratory, Isles of Shoals, in the Gulf of Maine. He and a colleague completely removed all the seaweeds from a one-meter strip extending down the rock surface from high tide to low tide level, and then followed the recovery process over several years. Shortly after clearing, the first seaweed colonizer to arrive was the green alga *Enteromorpha.* The next summer it had already been replaced by a luxuriant growth of *Fucus vesiculosus,* which persisted for more than fifteen years on parts of the strip where *Ascophyllum nodosum* had been the dominant seaweed prior to clearing.

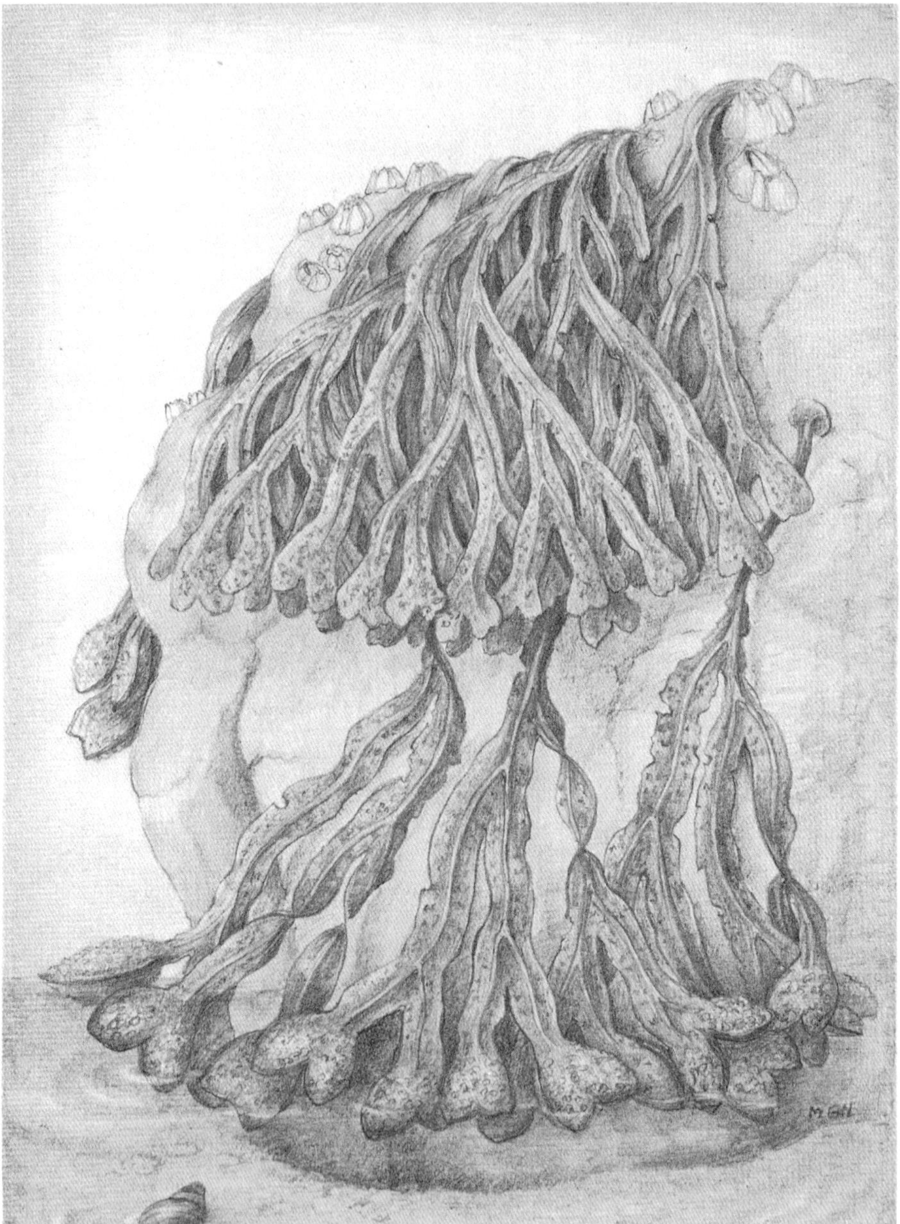

Fucus spiralis X 0.75

Fucus spiralis

Fucales

Derivation: from old Greek, meaning "seaweed."
Range: Long Island northward.

Except for the fact that *Fucus spiralis* never forms bladders, it is indistinguishable from *Fucus vesiculosus* in general appearance. The only sure method for distinguishing the two has to do with the microscopic nature of their reproductive structures—certainly not a handy method for use in the field (see the following discussion of *Fucus distichus*). However, two additional characteristics, though not infallible, are helpful in distinguishing *Fucus spiralis.* In the first place, *Fucus spiralis* is limited in distribution to the high intertidal zone. More is said about that below. Secondly, the swollen fruiting branch tips or receptacles, often look as though pressed out in a cheap mold. A ledge or ridge encircles the receptacle as though the mold edges did not quite come together. This is shown well in the illustration on the receptacle farthest to the left of the lower of the two plants. This characteristic is useful only with receptacles at full maturity and not beyond. Young receptacles of *Fucus vesiculosus* may show a similar ledge, and "going-by" receptacles of *Fucus spiralis* may lack it.

The relative distributions of *Fucus spiralis, Fucus vesiculosus, Ascophyllum nodosum,* and various kelps illustrate zonation quite well if you can find a place where all are present. One such place is at the Sandwich canal jetty. At the West Falmouth jetty you can find all but the kelps. Once you can separate the population of *Fucus spiralis* from that of *Fucus vesiculosus* by eye, you can determine that the former always occurs above the latter in the intertidal zone, and usually they do not mix appreciably. *Fucus spiralis* is confined to a zone that may represent only a few inches of vertical distance. *Fucus vesiculosus* ranges over a greater vertical height. *Ascophyllum* and *Fucus vesiculosus* intermix vertically, but the former is apt to be concentrated in the lower part of the *Fucus vesiculosus* distribution. The lower limit of the rockweeds (Order Fucales) is usually quite distinct and represents approximately the limit to which the tide recedes. This disjunction is made all the more obvious in areas where kelps occur, because it simultaneously represents the approximate upper limit of their distribution.

Living high on the shore, *Fucus spiralis* is exposed to the air for long periods while the tide is out. A danger during this time is that water will evaporate from its tissues causing damage to delicate cellular structures. Although *F. spiralis* lacks a waterproof covering as formed by land plants, it is able to tolerate prolonged periods of drying while still carrying out its basic cellular processes. Even when it has lost 85 percent of its water content, *F. spiralis* still shows some photosynthetic activity, and when resubmerged by the rising tide it recovers rapidly to its full capacity.

The various rockweeds fruit mostly during autumn, winter, and spring. The fruiting branch-tips of *Fucus* tend to persist on the plant for some time after they have ceased functioning. Those on *Ascophyllum,* however, are deciduous, being shed all at once in late spring, at which time they may come ashore in such quantities as to attract attention.

Fucus distichus X 0.75

Fucus distichus

Fucales

Derivation: from old Greek, meaning "seaweed."
Range: Virginia northward.

Fucus distichus is sparse on the south side of Cape Cod, but common on rocks between high and low tide marks from the northern side of the Cape northward. Plants are similar to *Fucus vesiculosus* in general appearance, but lack air bladders. The reproductive branch tips (receptacles) are quite different, however. In the more common *Fucus vesiculosus*, receptacles are swollen, soft to the touch, and only slightly elongated. In *Fucus distichus* they are scarcely thicker than the vegetative branches and firmer in substance. In the common *F. distichus* subspecies *edentatus*, the receptacles are several times longer than broad, often branching two or three times as shown here. (The upper plant is *F. vesiculosus*).

Fucus distichus occurs in several distinct subspecies. *F. distichus* ssp. *evanescens* has short receptacles (similar in length to *F. vesiculosus*, but flat), and there are two dwarf subspecies that are about four inches tall when mature.

Fucus and its relatives possess the most advanced life cycle and type of reproduction to be found among brown algae, although the cellular construction of the plant itself is not as complex as that of the kelps. The plant you see is the only free-living phase in the life cycle; there is no free-living microscopic phase. Its cells contain doubled chromosomes, so the plant is diploid. Generally among brown algae, diploid plants produce unilocular structures in which a reductive division takes place and from which haploid, directly-germinating swarmers (zoospores) are released. These initiate the separate, haploid, gamete-producing stage in the life cycle. Fusion of gametes reestablishes the diploid condition and gives rise to the new diploid plant. In *Fucus*, however, events are short circuited. The diploid plant produces eggs and motile male gametes (sperm) directly. These are formed within the cavities whose openings appear as dots scattered on the surface of the reproductive branch tips (receptacles). In *F. spiralis* and *F. distichus* both male and female organs are formed in the same cavity; while in *F. vesiculosus*, plants are of separate sex (a fundamental characteristic separating these species).

Chances for fertilization of the female gamete from one plant by a male gamete from a different plant (genetic intermixing), and for gametes finding partners in the vastness of ocean water around them, are enhanced by concentration in time and space. Gametes are are produced in the cavities in membrane-bounded packets. The drop of mucilage containing them is squeezed out of the mouth of the cavity by shrinkage of the receptacle tissue around it as the frond dries when exposed to air between high tides. Thus large numbers of matured but immobilized gametes slowly collect on the receptacle surfaces. When the tide returns, the first wave to inundate the receptacles dissolves the mucilage drops, bursts the packet membranes, and frees the gametes. Female gametes are large non-motile eggs. Male gametes are intensely active flagellated swarmers (sperm). They attach to the eggs, one effects fertilization with it, and the germling of a new *Fucus* plant is thus initiated.

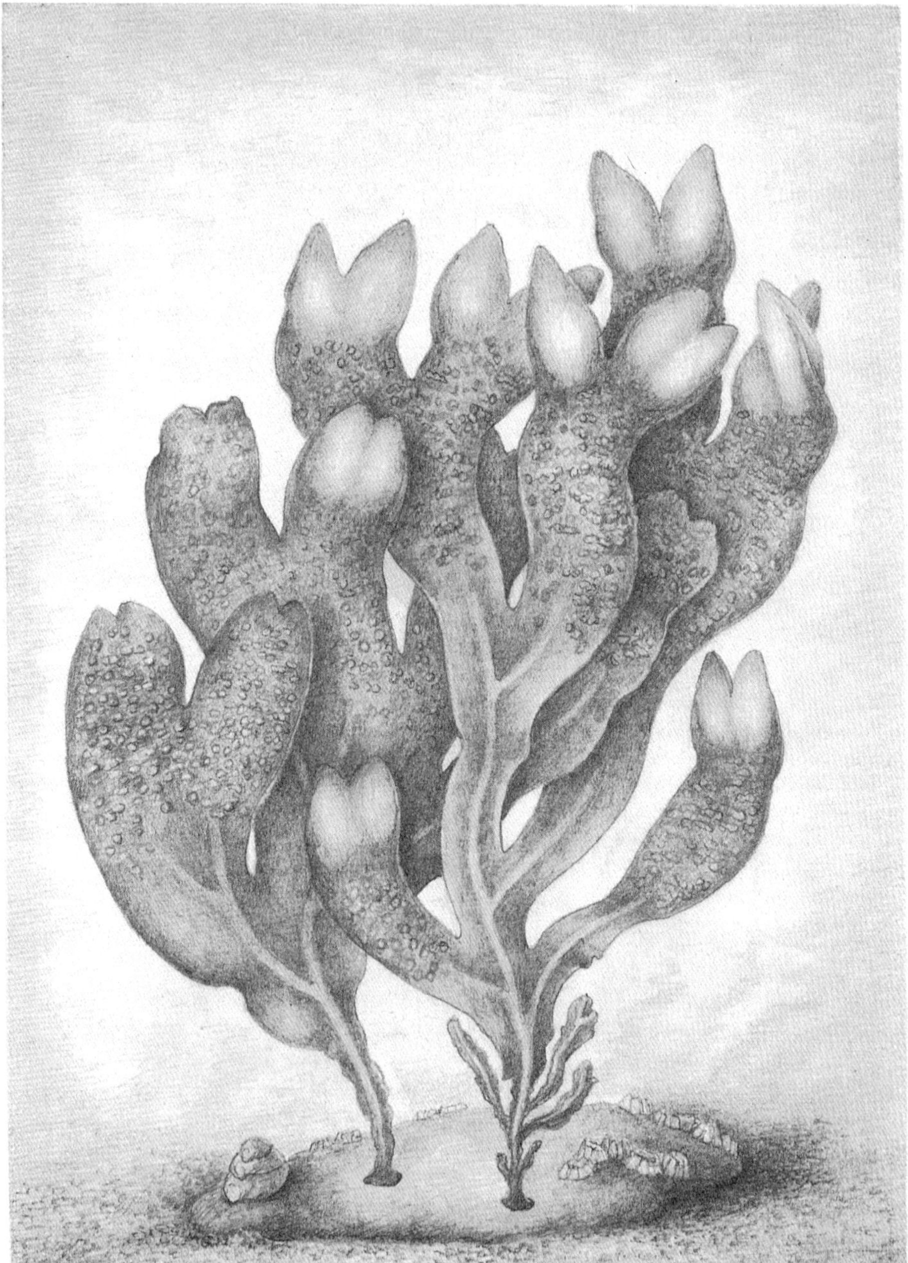

Fucus sp. X 1.0

Fucus species

Fucales

Derivation: from old Greek, meaning "seaweed."
Range: North Carolina northward.

The plant illustrated here is sterile or past its seasonal reproductive maturity and lacks air bladders. This makes it impossible to determine for sure whether it is *Fucus vesiculosus*, *Fucus spiralis*, or possibly *Fucus distichus*. The same would be true if one tried to identify germlings of almost any of the species of *Fucus*. They are quite alike, and no book yet describes the differentiating characteristics, if any.

Comparison of the three species of *Fucus* illustrated and discussed in this book, together with this sterile specimen, will help to give an idea of the characteristics used to separate species. Actually this thought is backward. The plants separate themselves into non-interbreeding populations of kinds which biologists then attempt to recognize and characterize. Sometimes they get frustrated in this endeavor, but, to put it in facetious words, the plants couldn't care less.

Sometimes when biologists think they have recognized separate and distinct species in nature, they find they can easily make them interbreed. Such a situation has happened among the species of *Fucus*, which can be made to produce hybrids with fundamental characteristics intermediate between those of the parent species. Although biologists use the species concept constantly and it means something quite definite to each person who specializes in studying a particular group of organisms, no one has been able to produce a definition of "species" that covers all circumstances of diversity found in nature. "Nature mocks at human categories."

That *Fucus vesiculosus* can produce air bladders is genetically determined. In other words, it is characteristic for this species alone. Whether a given plant actually will produce them depends on circumstances. In studying seaweeds on the West Falmouth jetty, for example, Kingsbury was able to find bladders only on those plants which either lived on the lee side of the jetty, or were located in the protected cracks and crevices between rocks on the windward side. Plants of the same populations located in fully exposed places always lacked air bladders. This makes sense when you realize that the bladders cause greatly increased resistance to water movement, or drag, about the plants and make them more apt to be torn off their moorings by storm waves.

It is always useful if possible, to distinguish among those characteristics which are genetically determined, those which are determined by environment, and those which combine both, in attempting to "understand" a species.

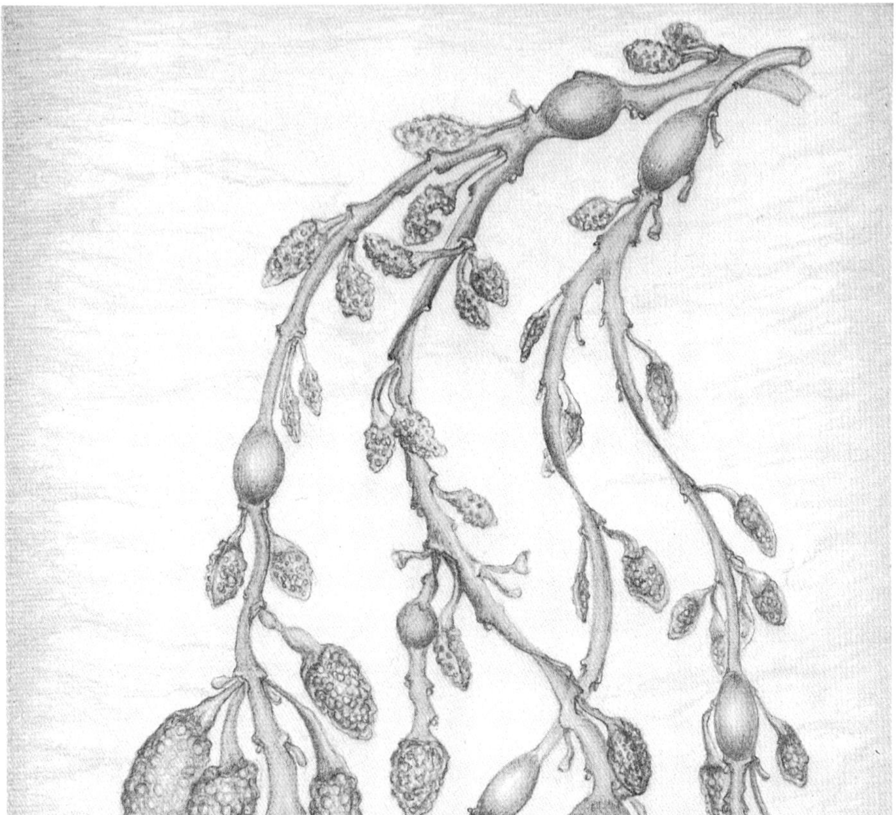

Ascophyllum nodosum -fruiting- X 1.0

Ascophyllum nodosum — Fucales

Derivation: "pod-leaved."
Range: North Carolina northward.

This very common rockweed inhabits the mid to lower intertidal levels. It can immediately be distinguished from other rockweeds *(Fucus)* by the absence of differentiated midrib and wings. Instead, the solid branches of *Ascophyllum* are round to oval in section, irregularly rebranched, and bear conspicuously swollen, thick-walled air bladders at intervals along their length. The illustration shows the terminal portions of two plants in early spring condition. The large, warty tips (mostly on short lateral branches) are fruiting receptacles, similar to those of *Fucus,* already well past their prime. The receptacles of *Fucus* are retained on the plant over a long period of time. Those of *Ascophyllum* appear in late winter, mature rapidly, and are promptly shed after release of gametes.

The branching pattern of *Ascophyllum* is not regular as in *Fucus,* and the major axes may bear numerous, short, thin lateral branches. Regeneration after wounding often gives rise to a cluster of branches of about equal size. The plant is anchored to the substratum by means of an irregular disc-like holdfast.

Where *Ascophyllum* lives, it often grows in such great abundance that it covers the rocks with a thick, treacherous carpet. The outer layers of such a covering dry in the air between high tides, but underneath, the branches remain wet and very slippery. Presence of a covering of *Ascophyllum* on rocks, or curtains of it hanging from projecting edges, provides shelter and protection from drying for a number of organisms which otherwise would be unable to exist in the intertidal zone. Your collecting is not complete until you have parted the curtains of *Ascophyllum* and examined the rock surfaces beneath for new finds.

If *Ascophyllum* is so common, why did it not immediately recolonize the experimentally scraped surfaces described on p. 113? Dr. Robert Vadas and his students at the University of Maine tested *Ascophyllum's* ability to recolonize by mixing gametes together to form fusion cells and pouring them on rocks. They were rarely able to attach to bare rock. This experiment confirmed that *Ascophyllum* is not particularly good at settling on a rock surface in moving water. Then why is it so common? First, it is unusually long-lived compared with most perennial seaweeds. If it is successful in settling, it will persist for eight to twelve years or even longer. Second, it is protected from herbivorous snail attack by the toughness of its tissues and by repellant chemicals. Subtle adaptations such as these can obviously lead to impressive consequences.

Artists Marcia and Edward Norman, who illustrated this book, were particularly impressed by *Ascophyllum.* The two illustrations on the following pages present additional characteristics of *Ascophyllum nodosum.* The first is of typical material in summer condition (sterile). The second shows a growth form (var. *scorpioides*) that is found in salt marshes along the coast. As illustrated, it is fruiting and also has somewhat more numerous air bladders than might typically be expected. These two drawings also emphasize the composition, form, and beauty to be found in even the most common of seaweeds.

Ascophyllum nodosum -sterile- X 1.0

Ascophyllum nodosum var scorpioides X 1.0

Sargassum filipendula X 0.50

Sargassum filipendula — Fucales

Derivation: from Spanish for "grape."
Range: tropics to Cape Cod.

The Sargasso Sea, has aroused a mystic fascination ever since it was first described by Christopher Columbus in his log of 1492. The simple coastal sailors who crewed the *Niña, Pinta,* and *Santa Maria* were tempted to mutiny because they feared rocks ahead (which fresh seaweed had always before meant), or entanglement in the windrows that *Sargassum* forms at sea. Today we recognize the Sargasso Sea as an intriguing region within the giant circulation pattern of the embracing North Atlantic. Few persons realize that *Sargassum* is the commonest subtidal brown seaweed in many places on Cape Cod.

Nearly a score of species of *Sargassum* have been described by botanists. Our particular Cape Cod species attaches to rocks or shells from just below low tide to considerable depths. The plants tend to be long and narrow, increasing in length with age, streaming erect in the water, buoyed up by their numerous spherical air bladders. The berry-like bladders and the dotted, leaf-like foliage, each blade of which has coarsely toothed edges and a distinct midrib, serve to identify *Sargassum* or "gulfweed" as it is sometimes called. Bladders may be few or entirely lacking in some plants.

An almost unfailing principle of construction among higher land plants is the relationship of a branch to the stem bearing it, and to the leaves on that stem. Branches grow from buds that originate in the upper angle between a leaf and the stem it is on. Sometimes the branching pattern of a *Sargassum* mimics that of a higher plant closely, with a branching system composed of stem, bladders, and reproductive structures seeming to arise in the upper angle between a blade and the stalk bearing it. Such appearances are not homologies, because the development of a *Sargassum* plant is entirely different from that of a higher plant, and the branching pattern of a *Sargassum* is probably more accurately traceable in evolution to the divergencies of a *Fucus.*

Other species of *Sargassum* are occasionally cast ashore on the south shores of Cape Cod and the Islands. Possibly originating in the Caribbean and brought northward by the Gulf Stream and onshore storms, such plants may travel great distances, for unlike most other seaweeds they may not die quickly after having been torn loose from their original attachment.

Plants in the Sargasso Sea are sexually sterile, reproducing themselves by growth and fragmentation. A student in a Sea Education Association (Woods Hole) course who spent two weeks in the Sargasso Sea described it thus: "One may sail for hours without seeing a single plant, while at other times great aggregations appear. If there is a wind blowing, long, loosely intertwined windrows of *Sargassum* may be seen. These windrows often extend as far as the eye can see. They always follow the direction of the wind. In calm weather, large isodiametric flotillas of seaweed are common. Most of the plants float on the surface of the water, being held up by numerous air bladders which are always present. Some may drift along several meters beneath the surface, particularly the heavier plants overgrown with epizoans (attached animals) which weight them down."

Bangia atropurpurea X 1.0

Bangia atropurpurea

Bangiales

Derivation: honors N. H. Bang, of Denmark, a botanist.
Range: entire East Coast.

Bangia atropurpurea forms patches of tufted growth on rocks and wood between tide lines, usually near the high tide mark. Each patch is composed of a dense population of unbranched, soft, slippery strings, usually two or three inches long; sometimes longer. These, except for color, are not unlike the populations of *Ulothrix flacca* that inhabit the same general locale. Young *Bangia* is somewhere between red-purple and black in color, but as it ages it tends to bleach, becoming brown-purple and even yellow-brown.

You should be able to determine with a hand lens that the strings are unbranched, but the cellular structure is too small for observation without a microscope. Each string starts out as a filament (a single chain of cells). Then further divisions take place longitudinally within each cell of the chain, especially in the upper portions of the filament, so that it becomes multicellular in three dimensions and somewhat like a pulled-out baseball bat in shape. Old plants are irregularly swollen in diameter. Populations of this species appear, grow, become scencscent, die, and disappear completely over a period of a few weeks in spring and early summer.

Bangia is classified among the most primitive of red algae. Like the green (see *Ulothrix*) and brown algae (see *Ectocarpus*), the red algae are built of cells with walls, a single nucleus (exceptions), and one or more chloroplasts in each cell. Pigmentation is more complicated than that of either browns or greens. The chloroplasts of red algae possess a third class of pigments in addition to the chlorophyll (green) and carotenoid (yellow to brown) of both the green and brown algae. The third class of pigments is the phycobilins, and they are chemically different from either of the other two classes. Phycobilins exist in different specific kinds which may differ considerably in color. The complement of phycobilin and other pigments in the chloroplasts of a given red alga may change as its growing conditions change. This means first, that red algae may display a greater variety and range of color than either of the other two major groups, and secondly, that the color of a given species may not always be exactly the same even under good growing conditions.

In all three groups of seaweeds, pigments contained in the chloroplasts are functional in the food-making process of photosynthesis as long as they receive light. The actual conversion of light energy to chemical energy is brought about by one particular type of chlorophyll which is found in all algae and higher plants. The other chloroplast pigments are called accessory pigments, transferring the light-energy they receive to the chemically functional chlorophyll, which then makes it available in the general metabolism of the cell.

Porphyra umbilicalis X 0.30

Porphyra umbilicalis Bangiales

Derivation: Greek name for a purple dye.
Range: Virginia northward.

This is an intertidal seaweed, usually attached to rocks or pilings, or other algae. When the tide goes out, each delicate blade dries and shrinks somewhat, often pulling taut against the underlying rocks or seaweeds. It looks like a piece of wax paper, except for the color. Under water the plants are soft, flexuous, and mostly upright, waving almost transparent sheets of color gently about in the current. Except for color, they look much like *Ulva*, or because they are so delicate in substance, perhaps it would be better to compare them with *Monostroma*. Looking at dried plants, you would believe it difficult to collect them without tearing them. But they are a lot tougher than they appear and it is surprising how firmly you must pull on the edges of a healthy plant to make it tear.

Cape Cod has several species of *Porphyra*, difficult to separate with certainty. Their blades vary in color from purple to brown and may be one or two cells in thickness. They usually are at their best vegetative condition in spring and early summer. The basic structure of the blade is built up by a process of diffuse cell division.

The life cycle of *Porphyra* involves both a conspicuous blade phase and a microscopic filamentous phase, called the conchocelis phase. When blades of *Porphyra* disappear during certain seasons of the year, the plant has actually turned itself into its alternative phase, which grows on mollusk shells.

Porphyra is an important food product in the Orient, where it is farmed by both the Japanese and Chinese. The microscopic conchocelis phase is maintained in culture and is stimulated to produce spores which attach to nets. At the right season, the nets are transferred to bays where they are strung horizontally either between poles or on floating rafts. After the blades have grown to a sufficient size, they are harvested and processed into sheets of *nori*, which are used for flavoring and as wrappers for *sushi*.

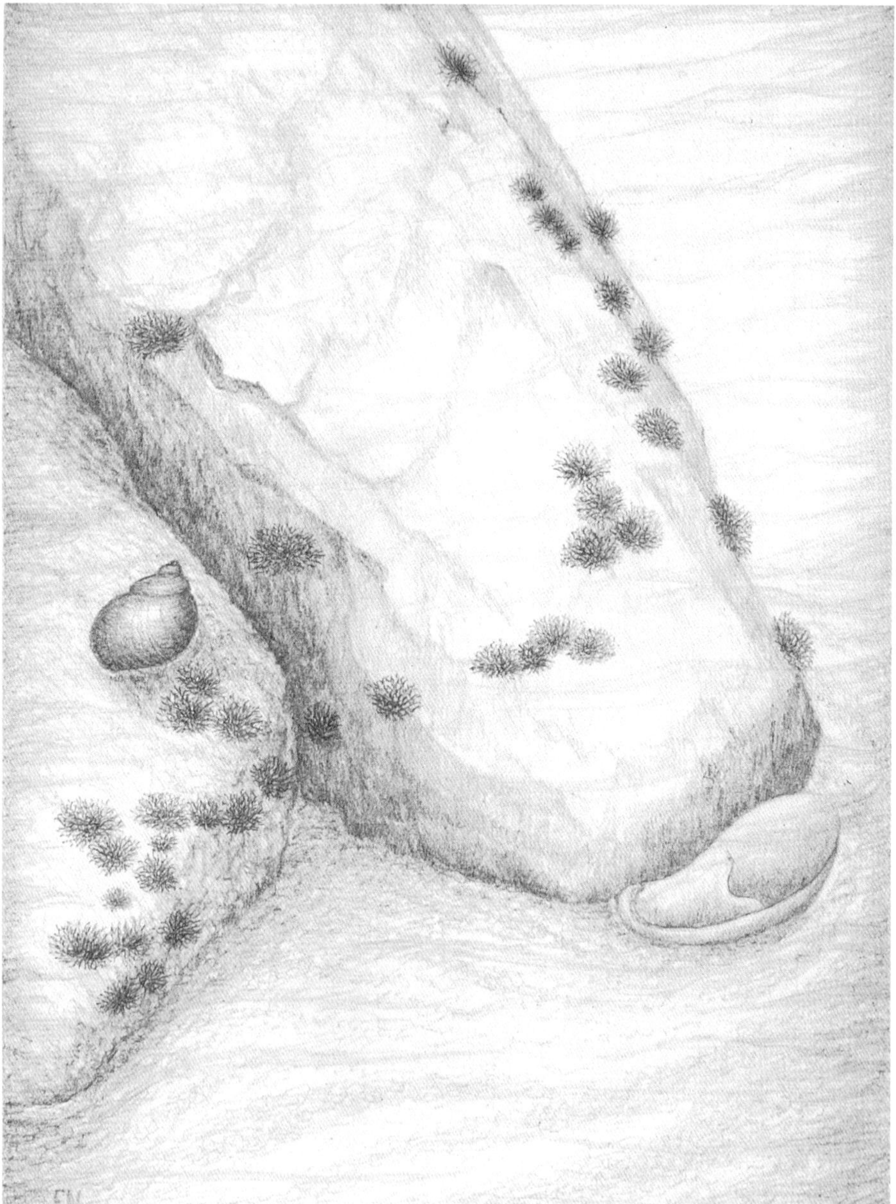

Audouinella purpurea X 1.0

Audouinella purpurea

Acrochaetiales

Derivation: honors J. V. Audoin, a French zoologist.
Range: Long Island northward.

This alga has been included as an example of a large and complex group of minute filamentous red algae which are impossible to identify without close attention with a microscope. They appear as tiny mounds of red, or red fringes, less than a half inch high, growing on seaweeds or rocks and pilings in the intertidal zone. Under the hand lens, it is just possible to discern that they are composed of clusters of branching filaments. Working out the individual genera and species with these general characteristics is complicated by a number of factors. First is their small size. Secondly, algae of similar appearance may belong to quite different groups with correspondingly different methods of reproduction. Finally, a thorough study of one of these filamentous plants often shows it is actually a stage in the life cycle of an already known, larger seaweed.

Audouinella purpurea is one of the largest of the primitive filamentous algae of the Order Acrochaetiales. Each tuft may be may be nearly a half inch high. It grows on rocks or pilings usually near or below low tide line (most of the others grow as epiphytes). It is found also on the holdfasts of *Laminaria* plants. Some of the other simple filamentous plants of the Acrochaetiales grow only in a single type of habitat, sometimes only on a particular species of host seaweed.

The number of species of red algae world-wide is as great as the number of green and brown seaweeds combined, and this relationship holds for our area as this book shows. Except for the most primitive members, exemplified in this book by *Bangia* and *Porphyra*, the red algae present great diversity of form and even greater complexity of reproduction. At the same time, they are nearly uniform in fundamental structure (interwoven filaments). Because the basic structure varies little, one must turn elsewhere for characteristics by which to recognize, separate, and organize these algae in some natural pattern of relationships.

Unfortunately, general appearance does not work. For example, simple filaments are found among the more primitive groups, such as the Bangiales and Acrochaetiales, as well as in the most advanced group, the Ceramiales. Scientists who specialize on studying red algae have sought order and pattern in details of the reproductive process and cellular structure, but the red algae present a reproductive situation as complex as in any group of organisms. Furthermore the necessary details are almost entirely microscopic and embedded three dimensionally in the solid tissues of the alga. Hence the red algae can be very frustrating, and only a few scientists may have studied important characteristics for a species, often in only one locality. Thus, information about a species in our area is often based on limited studies done somewhere else.

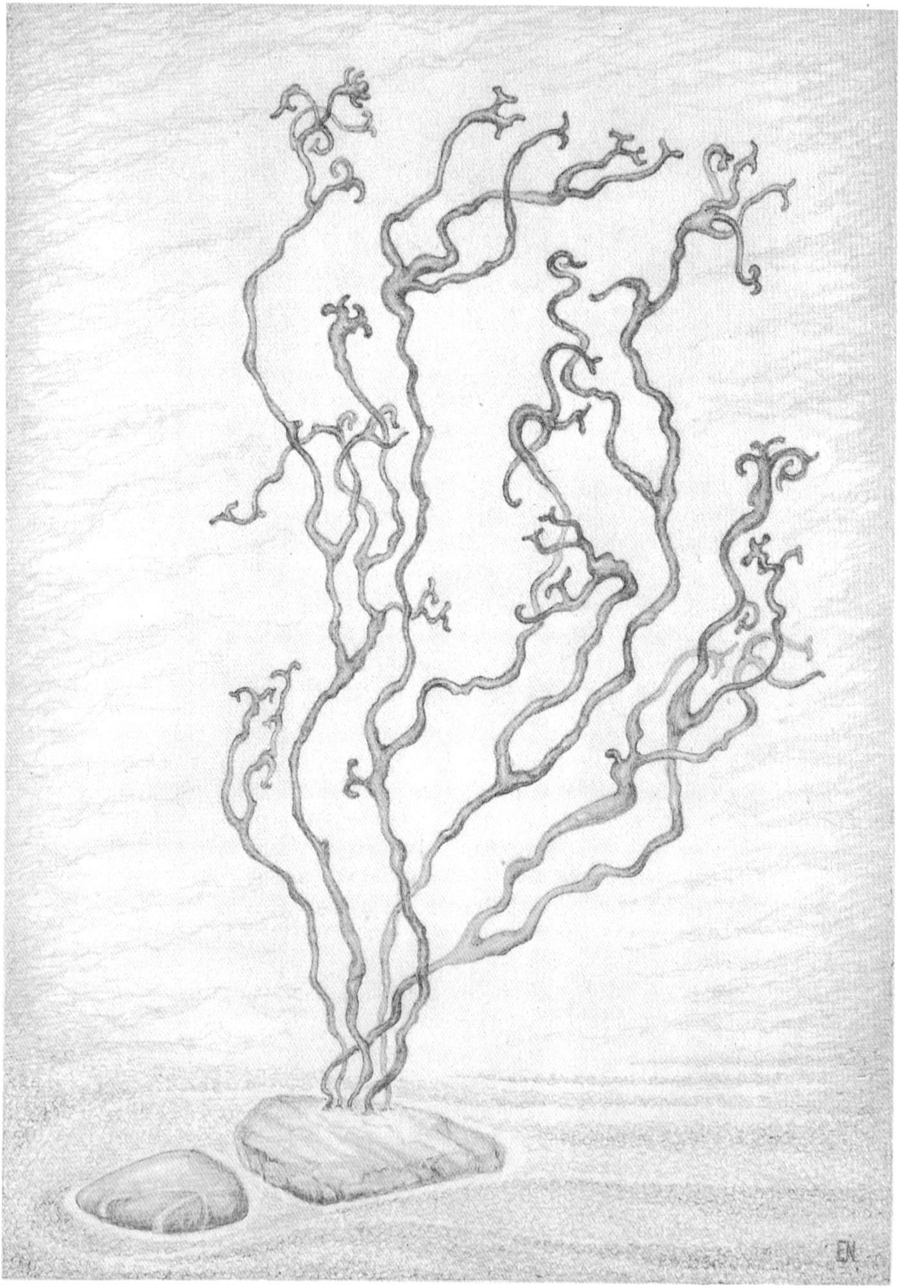

Nemalion helminthoides X 0.75

Nemalion helminthoides

Nemaliales

Derivation: "crop of threads."
Range: Long Island northward.

This is a characteristic mid-summer plant which forms extensive, slippery, almost slimy areas of growth on rocks, especially in exposed locations in the mid to low intertidal zone. Each plant is dull red or rusty brown, worm-like, from one to a few inches long, soft, and slippery. Some plants of the population will be without branches. The majority are branched irregularly one or more times. The diameter of the axis varies, and its surface is indistinct.

The structure of *Nemalion* is similar to that of *Eudesme* of the browns and because of its looseness, is similarly useful in demonstrating the basic structure of many of the red seaweeds. Each axis is composed of a medulla of loosely intertwined, mostly colorless microscopic filaments, and an indistinct cortex of outwardly turned filament-branch-clusters arising from them. The surface of the plant is made up of the tips of these radial filament branches projecting to more or less the same level. Growth in length of each axis occurs through divisions of cells located at the tips of several of the axial filaments. The cells themselves are not barrel-shaped or cylindrical as in the filaments of most brown and green algae. Instead each filament, as seen under the microscope, looks more like a string of eggs. The reason for this, in part, is that the cell walls are almost invisible having nearly the same refractive index as the water in which they are mounted, and what you see is only the contents of each cell. This is true for most of the red algae beyond the Order Bangiales.

In its typical form this seaweed, as you collect it, contains only one complement of chromosomes per nucleus (per cell), and hence is haploid. It reproduces sexually. Male gametes are produced as single-celled, bud-like outgrowths at the tips of cortical filaments and when ripe are simply shed into the water. They lack flagella (whip-like structures giving motility) and are carried passively by the water. In the red algae, female gametes are never released from the cell producing them. Male gametes are termed spermatia (singular: spermatium), and the cell bearing the female gamete is termed the carpogonium. Carpogonia are formed singly at the tips of few-celled specialized filaments scattered in the outer parts of the cortex. Each carpogonium has a terminal finger-like projection. If a spermatium touches on this, it will stick, the walls between them dissolve, and the male nucleus passes down the interior of the projection to the female nucleus in the carpogonium at its base. Fusion of nuclei, the essential sexual act, then takes place to produce a cell with a double complement of chromosomes. This diploid cell proliferates, growing into a number of new filaments composed of diploid cells. These filaments grow forth from the embedded carpogonium forming a dense spherical cluster which may be seen by hand lens as a dark speck or "birds nest" within the cortex of a fruiting alga. The terminal cells of this cluster are shed as spores which develop into a microscopic diploid filamentous stage resembling a small *Audouinella*.

Bonnemaisonia hamifera X 1.0

Bonnemaisonia hamifera

Bonnemaisoniales

Derivation: named after T. Bonnemaison, a French botanist.
Range: Long Island northward.

Bonnemaisonia was first found on American shores at Nobska Point, to the east of Woods Hole, in 1927. Probably introduced from Europe, it has since become well established from Cape Cod to Long Island.

Bonnemaisonia is usually found as tangles of bright or rose red, bushy plants in the wash. It grows below low tide to considerable depth. The major axes are of small diameter and are densely clothed with yet smaller branches. The ultimate branchlets are very numerous, short, and very delicate. The internal structure is similar to that of *Nemalion* but much more compacted.

Bonnemaisonia possesses a peculiarity of structure by which it can be positively identified. Among the delicate branchlets in the outer part of the branching systems, look for greatly enlarged, hook- or scimitar-shaped branches, each about a quarter inch long if uncoiled, and of diameter as great as or greater than that of a major axis at the bottom of the plant. These swollen hook branches taper at the tip and end in a sharp point. They may be only slightly coiled, or coiled in a complete circle. Of our flora, *Hypnea* is the only other red seaweed with similar coiled branches, but its general characteristics are quite different and the two are not easily confused.

The life history of *Bonnemaisonia* is similar to *Nemalion*. The carpospores produced at the tips of the "bird's-nest" filaments that arise from the carpogonium after fertilization grow into a filamentous alga previously recognized as *Trailliella intricata*. Another type of spore (the tetraspore) is produced by the *Trailliella* phase, and it is these spores from which more of the *Bonnemaisonia* phase grows.

Trailliella phase on Chondrus X 1.0 *enlarged detail*

The first time *Trailliella* was collected on American shores was the same year that *Bonnemaisonia* was first found here. *Trailliella* grows in a tuft which is usually between one half inch and an inch in diameter, and deep rose red in color. It grows on a variety of other seaweeds at low tide and just below. The branching pattern is alternate to irregular, and special tiny colorless cells are present between the regular cells of filaments but require a microscope to be seen.

At this point, it is useful to look at the life cycle of a typical red alga, as summarized in the diagram below. Most red algae have three phases—a sexually reproducing haploid phase (gametophyte) often with separate male and female individuals, a diploid phase (carposporophyte) that develops as a "bird's nest" in (or on) the haploid phase, and an independent diploid phase (tetrasporophyte). The two independent phases may be identical in general form, or they may differ drastically.in appearance. *Bonnemaisonia* is among the latter.

In the life cycle of *Bonnemaisonia*, the independent haploid and diploid phases are very distinct entities, so much so that scientists originally gave them different species names. Now that both entities are linked as parts of one life cycle, the name *Bonnemaisonia* is used as having come first. *Trailliella* no longer has validity as an official name but is used informally to refer to the diploid stage of *Bonnemaisonia*. A similar situation occurs in *Nemalion* , except that the independent diploid phase is a microscopic filament. In other red algae as we shall see, both haploid gametophytes and diploid tetrasporophytes often have a similar or grossly identical appearance.

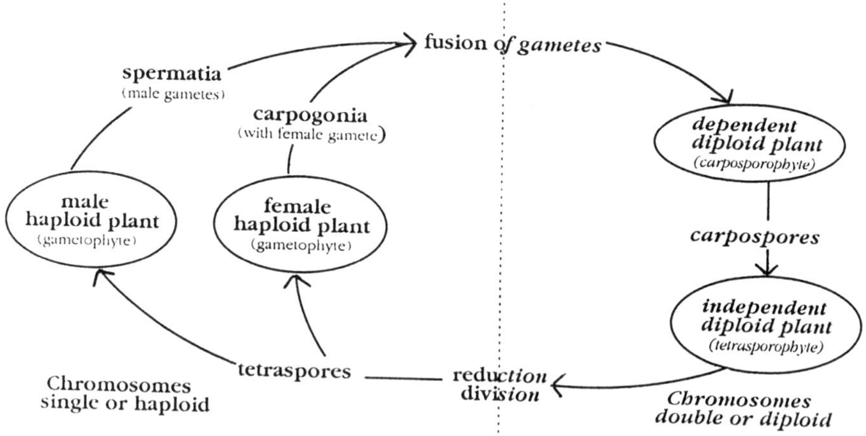

LIFE CYCLE OF A TYPICAL RED ALGA

The Fundamental Classification of Red Algae

Classificational systems for living organisms must be based on stable genetic characteristics, uninfluenced by the environment. In the red algae (except Bangiales), the stereotyped basic filamentous construction has not been useful in classifying red algae into major groups, nor has general form provided a useful basis for a taxonomic system, especially when there is a great deal of variation in overall appearance (see, for example, *Chondrus)* or a major difference in the form of the life-cycle phases (see *Mastocarpus, Ahnfeltia, Bonnemaisonia*).

Unable to use such characteristics, scientists studying red algae turned to details of the life cycle, especially related to sexual reproduction and the development of the carposporophyte phase. This system was elaborated and refined by the Swedish phycologist Harald Kylin during the early part of this century before it was possible to grow most red algae in culture. Culturing has since permitted scientists to vary growth conditions while looking for stable characteristics, to study the reproductive processes more carefully, and to identify alternate stages with different forms in the life cycles of many red algae.

More recently scientists have used electron microscopes to examine cells in great detail and have identified certain cellular structures that appear to be particularly useful in classifying algae. In particular, the Orders (the names ending in "-ales") of red algae, earlier based on their lifecycle details, are now being reevaluated based on characteristics of the plug filling the pore between adjacent cells (see *Ahnfeltia*). The result is seen in many of the changes in the classification of red algae since the first edition of this book. To give each Order a consistent pit structure, some formerly recognized Orders have been combined, while others have been separated into two or more new ones.

While we like to think of taxonomic classification as a fixed framework on which to organize species into a logical system, this framework is subject to revision as more is learned about different algae and their evolutionary relationships, and new techniques are developed for examining them. For instance, a new approach to determining the relationships among algae uses molecular techniques, similar to the genetic fingerprinting used by forensic scientists, to "read" segments (genes) of the DNA in the nucleus of cells. The degree of similarity of the "text" of these segments indicates the degree of relationship among different species. As this technique is applied to red algae, there will no doubt be further fundamental changes in how we classify them.

Gelidium pusillum X 1.0

Gelidium pusillum

Gelidiales

Derivation: "congealed."
Range: East Coast north to the St. Lawrence River.

This is our only species of *Gelidium*. Many species of this genus grow along American and Oriental shores of the Pacific, and they are mostly larger plants. Some of them are harvested commercially, especially in Asia and Latin America, and serve as the basis of the agar industry. Economic uses of red algae are discussed at *Chondrus*.

Gelidium pusillum is not conspicuous and it is not common in Cape Cod waters. Plants are gregarious, forming open or tight tufts mostly one to two inches tall. The stalks and branches are firm, almost wiry, and hardly greater in diameter than that of a coarse thread. Young plants are dull red or black-purple, but may bleach to a brown-yellow under some circumstances of growth. The best clue to identification is the nature of the short lateral branches in the upper parts of some or all main branches. They often occur in more or less regular pinnate arrangement so that the tips of the plants look like tiny Christmas trees pressed out flat. The plant shown here is embedded in a mass of "sea pork" (*Amaroucium*, a compound Ascidian animal, or "sea squirt").

In the life cycle of *Gelidium*, sporebearing filaments (of the embedded carposporophyte) develop from a carpogonium following fertilization as in other red algae, but unlike the "bird's nest" cluster in *Nemalion* , the carposporophyte filaments in *Gelidium* wind among each other to form a very complex spore-producing mass within the tissues of the parent plant. This mass produces a definite bump in the axis where it is located. When ripe, the bump opens by two holes, one on either side of the branch, opposite each other, through which the carpospores are released.

The carpospores generate a plant which looks exactly like the sexual stage but is diploid. At maturity, the diploid plant produces spores in groups of four at certain locations in its tissues. As the result of reduction division, these are again haploid and grow into the haploid, sexual plants. Because they are produced in fours, they are called tetraspores, and the type of plant bearing them is called the tetrasporophyte.

You can't tell which type of plant (haploid sexual or diploid tetrasporophyte) you have collected from gross characteristics alone. Only the presence of female "bumps" or following with a microscope what happens to the reproductive cells will serve to differentiate between these two kinds of free-living plants as you find them in nature.

Palmaria palmata X 0.50

Palmaria palmata

Palmariales

Derivation: "palm-like."
Range: New Jersey northward.

This is the seaweed called "dulse." For generations it has been eaten in Scotland, Iceland, and Ireland. The practice came to this country with settlers and has supported a small but steady market for dried dulse. At present much of the dulse in markets on the East Coast comes from Grand Manan Island. Dried or fresh dulse can be chewed like gum. It has a flavor that is perhaps best described as marine, and for most persons it seems to be an acquired taste.

Palmaria is more common farther north than at Cape Cod. Here it is found chiefly below low water, and is often heavily burdened with epiphytes (not suitable for eating). The plant is perennial, growing throughout the year. A typical plant looks something like a mitten, imperfectly formed for a giant hand. The round stalk is very short and broadens gradually into the large, flat, fairly thick blade. Plants are six inches to a foot or a little more in length, and up to six inches or so wide. As the blade expands upward in most specimens it divides into two or more equal or unequal sections. These may redivide, but the ultimate divisions remain comparatively broad and blunt, like large lobes. Some plants bear many small blades from the edge of the main blades, almost as though new plants were growing from the old one. The upper parts of a growing plant are thinner and lighter in color than the older parts at the bottom, where the color is deep red or almost black. The smaller plant in this illustration is *Callophyllis.*

Palmaria and its close relatives (in the Palmariales) are "odd balls" among the red algae. They have a distinctive life cycle that has been recognized only relatively recently. For a long time, field biologists could only find haploid male plants and diploid plants (tetrasporophhytes), both of which appeared to behave normally in their reproductive activities. Where were the females?

When, finally, *Palmaria* was grown in culture the diploid plants produced spores (haploid tetraspores) as in other red algae, and these spores developed into haploid male and haploid female plants. The mature male plants were typical for the species in size and appearance, but it was soon discovered that the previously missing females were tiny (less than a sixteenth of an inch) and inevitably overlooked in nature. The females matured rapidly. When fertilized by gametes (spermatia) from a male plant, the females produced the diploid phase directly from the female reproductive structure (carpogonium), and the massive diploid phase (the tetrasporophyte) eventually overgrew the tiny female completely. The carposporophyte and carpospores had been eliminted from the life cycle.

How were the details of this life cycle originally observed? John van der Meer and Edna Todd were very fortunate to work with a color mutant of *Palmaria* displaying a normal red color in the diploid condition but distinctly green in the haploid condition because of a genetic defect. Thus van der Meer and Todd were able to distinguish readily the tissues of the tiny haploid female plant (green) from the tissues of the massive tetrasporophyte (red) developing from it.

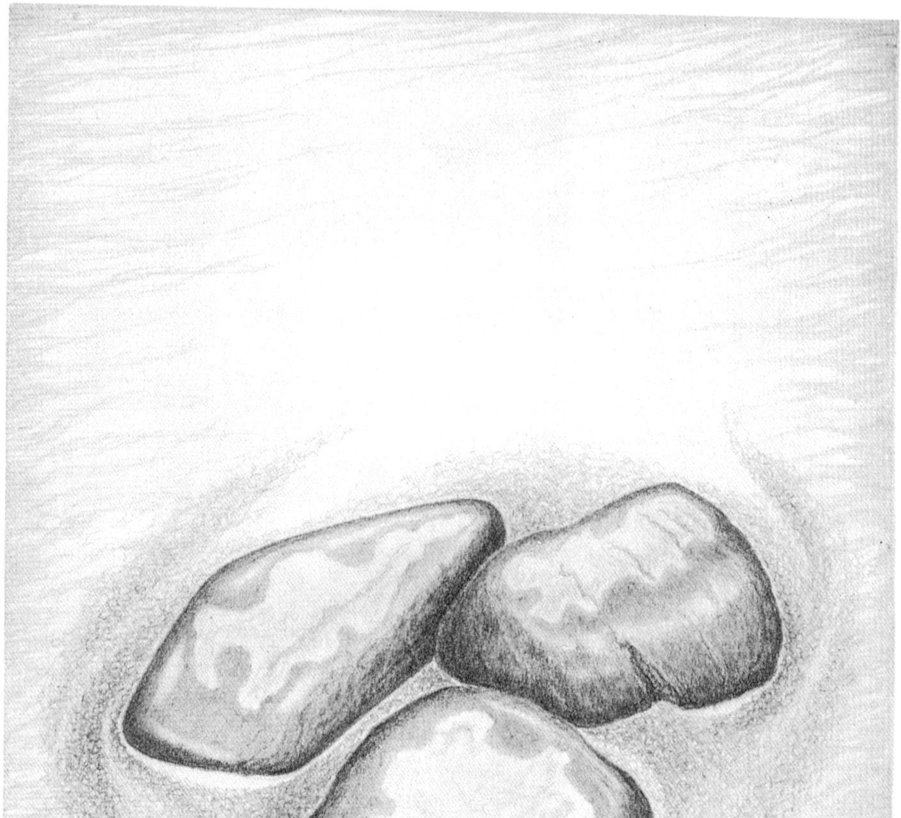

Hildenbrandia rubra X 1.0

Hildenbrandia rubra

Hildenbrandiales

Derivation: honors F. E. Hildenbrandt, professor at Vienna.
Range: entire East Coast.

This is among the commonest of red algae, yet it is perhaps the most easily overlooked. Except for its color, *Hildenbrandia* looks almost the same as the rock it is growing on, and in some parts of the Cape where pink granites are common, even the color is not conspicuously different. *Hildenbrandia* forms a paper thin covering of varying extent and irregular outline on hard surfaces near and below low tide. These are bright to rusty red, becoming darker with age. Their most striking characteristic is the tenacity with which they are attached to the rock surface. The plant layer follows the microscopic ups and downs of the rock surface and is so firmly attached at all points that you are as apt to get rock powder as plant tissue when you attempt to scrape it off with a knife.

A mature *Hildenbrandia* is several cells thick. A single layer of cells initially grows over the rock surface. From these cells, short filaments grow upward, and closely pressed together, form the substance of the film. In this type of construction, *Hildenbrandia* is not unlike *Ralfsia* of the browns, and sometimes it is difficult to tell the two apart. Young *Ralfsia* may be as thin as *Hildenbrandia*, and a rusty brown is not unlike a rusty red.

Hildenbrandia is the commonest of the encrusting, uncalcified red algae. Several others occur, however, especially to the north of Cape Cod and are distinguished on the basis of characteristics such as location and organization of reproductive structures. These things are not easy to determine in an organism which resists collection as effectively as these do.

The only practical way to collect a *Hildenbrandia* is to collect the whole rock, unless you have a geologist's hammer along. Finding a rock small enough to collect which has *Hildenbrandia* (or any other seaweed) on it is not easy for a very good reason. The smaller the rock, the more apt it is to be rolled around by the water. Not only may this grind off whatever had attached to it, but also what was top may become bottom and so shaded that plants, which need light, cannot grow there. When you collect on a warm summer day, you must exercise imagination to picture the same place in a winter storm. It is the latter, not the former, which will determine whether a stone rolls, and a rolling stone retains no seaweed.

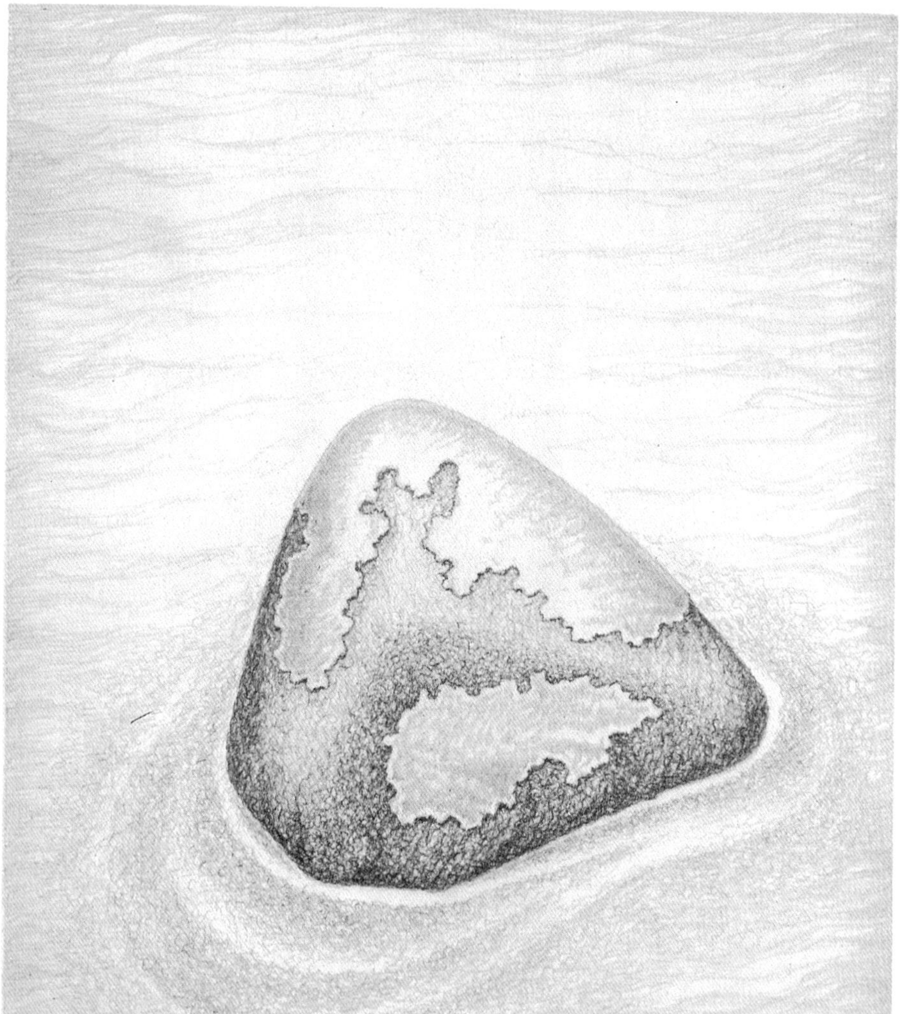

Phymatolithon lenormandii X 1.0

Phymatolithon lenormandii

Corallinales

Derivation: "rock-swelling"
Range: Cape Cod northward

This is a representative of a group of encrusting red algae which secrete a material akin to limestone. These algae, called encrusting corallines, are hard, rock-like crusts of irregular outline, adhering to the surfaces of stone and to other seaweeds. *Phymatolithon lenormandii* commonly grows under *Ascophyllum* or *Fucus* in the intertidal region. The encrusting corallines are lavender-white in color, and may be marked by concentric zones of slightly differing intensity of color. Larger crusts may be an inch or more across and of appreciable thickness. Dead plants are bone white. Each crust is irregularly round in outline if able to grow equally on all sides, but they often appear fan-shaped or entirely irregular because of injury, competition, or the nature of the surface on which they are growing. The growing edge is usually distinct, may appear as a series of broad crescents, and may be slightly upturned in old plants. Internal structure is much like that of *Hildenbrandia*, with lime deposited around the cells

Several other encrusting corallines grow in tidepools. Examples are *Phymatolithon laevigatum* with a rough surface and *Clathromorphum circumscriptum* with a smooth surface. Smaller, thinner crusts are often on the surfaces of eelgrass or coarser seaweeds. Perhaps the commonest of these is *Melobesia membranacea*. Encrusting algae must not be confused with crusts of ectoprocts or bryozoans which are neither round in outline nor concentrically marked. Occasionally in Cape Cod waters, and commonly farther north, very thick corallines are found, with distinct lumpy or knobby appearances. A common representative of these is *Lithothamnion graciale* .

Encrusting corallines occur in coastal waters throughout the oceans. They are important components of coral reefs in tropical waters where they contribute to the carbonate structure of a reef and help bind the reef together. Their rock-like form protects them against herbivorous animals such as sea urchins, but they are relatively slow growing compared to other seaweeds and may become overgrown by rapidly growing species.

Some encrusting corallines hold the record as the deepest growing seaweeds at 880 feet in very clear water near the Bahamas. In the Gulf of Maine, encrusting corallines are found on the sea bottom at 200 feet where less than one tenth of one percent of the light at the sea surface reaches them.

Encrusting, lime-precipitating algae are particularly difficult to study not only because of the difficulty of collecting them intact from their substrate, but especially because one must treat them with chemicals to dissolve the lime before their internal structure can be properly examined. Consequently, only a few persons have looked at them closely.

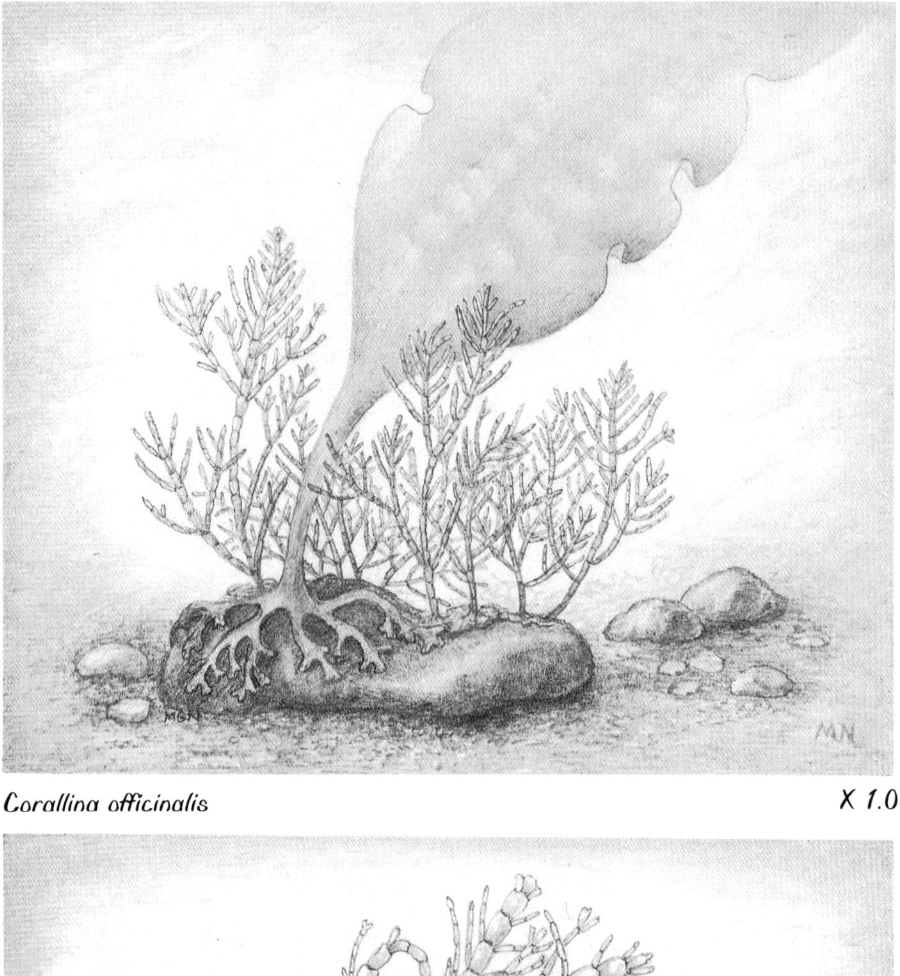

Corallina officinalis X 1.0

Corallina officinalis enlarged detail

Corallina officinalis

Corallinales

Derivation: "coral-like."
Range: entire East Coast.

Corallina officinalis is the only upright-growing, lime-precipitating alga in our area and identification, therefore, is easy. Each plant is composed of upwardly branching axes that are divided into segments. The segments are rigid and may be cylindrical or slightly fan-shaped. The joints between segments are flexible zones of tissue where no lime is precipitated, allowing the plants to bend and give in the waves that might otherwise shatter them. *Corallina* can be recognized immediately by these characteristics.

The "lime" of *Corallina* and related seaweeds is composed mostly of calcium carbonate precipitated from the water onto the surface of cells in a crystalline form called calcite. No one knows why some seaweeds form hard outer coverings in this way and why others do not. A thick layer of limestone on the outside is a considerable disadvantage to an organism which must get sunlight to live. How calcareous algae are able to compete with others whose photosynthetic abilities are not hindered by a light-shade on the surface is possibly associated with compensatory advantages of an exoskeleton. A hard calcareous layer may protect against damage from herbivorous animals.

Calcium carbonate, unlike most precipitating salts, becomes more soluble in water as the temperature *decreases*. Tropical marine waters are saturated or supersaturated with carbonates, while marine waters near the poles are less than ninety percent saturated with carbonates. Being limey is easy, almost unavoidable, in the tropics, but up north secreting calcium carbonate from seawater requires some effort. Certain coralline algae have found a way, at the cost of energy, to secrete lime from these waters. The next time you see a *Corallina*, consider that this alga is working to pull lime out of our northern waters, is controlling lime deposition in zones in its tissues leaving flexible joints between, and is coping with having thus reduced its ability to receive light for photosynthesis. Quite a plant!

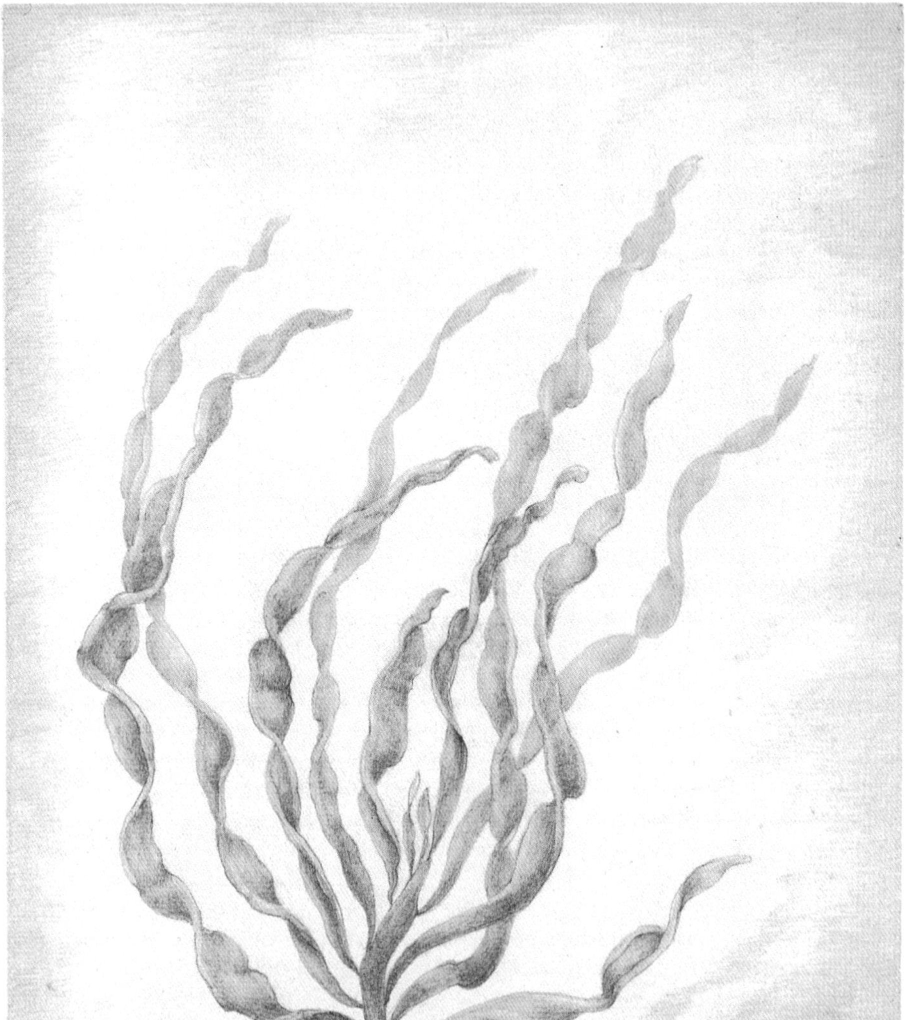

Dumontia contorta X 1.0

Dumontia contorta Gigartinales

Derivation: honors G. L. M. Dumont de Courset, a French botanist.
Range: Long Island northward.

This species is a common seaweed of northern Europe and Alaska, but had not been collected on the East Coast of North America until found in a tide pool at Harpswell, Maine in 1913. Since these tide pools had been examined rather closely at the same time of year in 1909 and earlier, without yielding *Dumontia* the plant must have been introduced during the intervening four year period. It was found at Kittery Point, extreme southern Maine in 1914, and had reached the Buzzards Bay area of Cape Cod by 1928, at which time it was reported to occur in abundance at Cuttyhunk and Penikese Islands, Sconticut Point off New Bedford Harbor, and at Woods Hole. These observations are not adequate by themselves to establish with certainty the appearance and spread of *Dumontia,* but they suggest a rate of spread similar to that for *Codium,* introduced more recently. *Dumontia* now may be found from Long Island to Nova Scotia and farther north. This distribution correlates with that of many seaweeds and suggests that *Dumontia* has spread as far as it can under our East Coast conditions.

Dumontia grows on stones and shells just below low water, but never in places where it may be exposed to prolonged drying. The plants usually appear in groups. Each is between one half and one foot or a little more in length and dull red in color. Old plants, especially those exposed to intense light, turn yellowish, particularly in the upper parts of the branches. Each plant has an obvious, solid axis, attached to the substrate by a small pad. Numerous large branches are inserted more or less randomly on the main axis, and rarely rebranch. The most conspicuous characteristic of *Dumontia* is the fact that the major branches are distinctly larger in diameter than the axes bearing them (but see also *Agardhiella)* and generally at least equal to the main axis in length. Older branches become hollow and may be inflated or squeezed flat. If the latter, their edges usually remain rounded as shown in this illustration. The branch tips commonly decay before the rest of the plant. *Dumontia* is in its best condition with us between May and July, disappearing later in the summer.

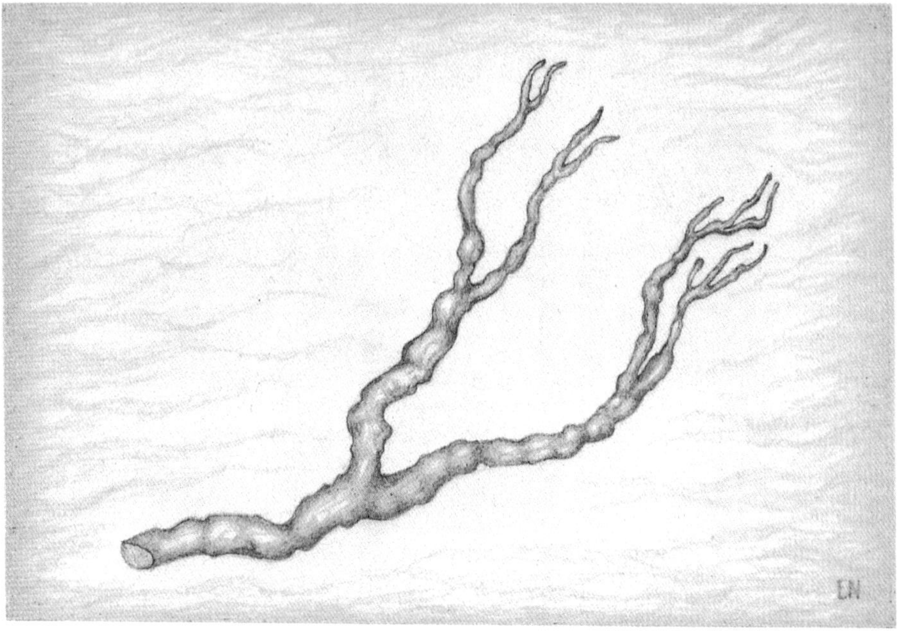

Platoma bairdii X 1.0

Platoma bairdii Gigartinales

Derivation: "oar-bladed."
Range: Massachusetts to Nova Scotia.

Back when a course devoted to the algae was taught at the Marine Biological Laboratory, casual visitors to the top of the cliffs at Gay Head on Marthas Vineyard, admiring the spectacular colors of clay, might wonder what was going on when the laboratory boat unloaded a class of some 25 students onto the narrow shore below them. As a reader of this book can guess, these people with plastic buckets, wading at the edge of the water and snatching at things floating by, or walking with head nearly between their knees examining the line of wash deposited along shore, were looking for rare and unusual seaweeds. Usually they were having a contest to see who, if anyone, could find some *Platoma.* This seaweed is entirely nondescript and esthetically uninteresting. Its interest is in its rarity. Dr. William Randolph Taylor described it as "apparently a plant of the lower littoral, growing on stones; very rare in this country", and known only from southern Massachusetts. Professor Farlow knew of only one collection of it prior to 1881, and that was at Gay Head. During the years classes went to Gay Head, one or a few pieces were generally found. Subsequently, *Platoma* has been reported from the Canadian coast and from Europe but information on this alga is very limited.

The fragments that wash up to Marthas Vineyard are rarely whole plants. Instead they are pieces an inch or two long. The plant is a very broad, delicate-walled cylinder, usually somewhat or mostly flattened. The main part may divide irregularly into one or two branches alike in general appearance. Except for the branching and rosy color, it is not unlike a small piece of a well-developed *Enteromorpha intestinalis.*

The general public finds it difficult to believe that people with buckets at the sea-shore are not finding something to eat. Shellfish wardens display a similar reaction. If you intend to collect seaweeds along our shores, you should be prepared to explain this eccentricity convincingly to an unbeliever. They tell the story of a Harvard professor who once was seeking rare fungi that grow only on goose dung. With basket in hand he followed a flock of geese across a field that happened to be near a mental hospital. He ended up in the hospital and had great difficulty getting out. Sometimes it is simpler, we have found, to concoct a story about something in your bucket being edible than to try to tell the truth to a persistent skeptic.

On the other hand, you may be accosted by some persons, children particularly, who take you at your word and display considerable interest in learning something about what you are doing and what you are finding. It's fun to talk with them.

Gloiosiphonia capillaris X 1.0

Gloiosiphonia capillaris

Gigartinales

Derivation: "sticky tube."
Range: Long Island Sound northward.

This alga is rare, but in the few spots where it occurs, there is usually a good population. The large, rosy-red, soft, bushy plants grow on rocks and shells, usually just below low tide or in tide pools. They appear in early spring and are gone by mid summer. To distinguish *Gloiosiphonia* from other red seaweeds of similar appearance, note the following. There may be one or a few main axes, each light-colored, cylindrical, hollow toward the base, extending nearly the height of the plant, and tapering at both ends. The main axis begins to bear lateral branches an inch or so above the base. These are mostly horizontal, randomly inserted on the stalk, tapered abruptly at the base and less so toward the tip. They in turn bear smaller but otherwise similar branches. The branching systems borne on the major lateral branches resemble individual bushes. The ultimate branchlets are very fine and threadlike. Some plants in fruiting condition bear warty swellings in the branches.

Gloiosiphonia can be used as an example of the complicated reproduction characteristic of the Order Gigartinales. The plant you collect may be either haploid or diploid. The two types are vegetatively indistinguishable. The basic plan of construction as seen by microscope, is like that described for *Nemalion*, except that the filaments of the cortex are more firmly knit together and the loose medulla tends to pull apart, becoming hollow with age. The haploid plant is the sexual stage in the life cycle. The spermatia (non-motile male gametes) are similar to those of *Nemalion*. The female reproductive apparatus consists of a specialized filament of the cortex. At the tip of the female filament is produced the female cell or carpogonium, with its characteristic finger-like projection to the surface of the plant. If a spermatium drifting passively in the currents happens to touch the carpogonial projection, it sticks. Adjoining cell walls break down and the male nucleus (haploid) enters the carpogonium and migrates through the protoplasm to the vicinity of the female nucleus in the base of the carpogonium. Fertilization ensues (nuclear fusion) and establishes the diploid condition of the chromosomes. What happens after fusion is described under *Callophyllis*, the next alga.

Callophyllis cristata X 1.0

Callophyllis cristata

Gigartinales

Derivation: "beautiful leaf."
Range: New Jersey northward.

Some lucky day, when you are walking the beaches after a storm, you may find pieces of this spectacular alga washed ashore. In color and form it is one of the most esthetically pleasing of all East Coast seaweeds.

Callophyllis crystata is rare south of Cape Cod, more common northward. Look for it in places where deep, cold water is near the coast, as at Gay Head on Marthas Vineyard. The seaweed grows at some depth and you must depend on specimens washed ashore if you do not dive. Look particularly at the kelps on the beach after a storm, for *Callophyllis* is commonly found attached to their holdfasts.

Each plant consists of a single, much divided blade, usually one to three inches tall. All parts of the plant are flat. Branching is irregularly fan-like. The width of the branch may vary along its length. The ultimate branches are quite delicate and the entire plant has much the effect of coarse lace. Under the hand lens, each branch can be seen to end in two or more blunt, finger-like lobes. Reproductive plants may bear large numbers of tiny, berry-like swellings on the edges of the branches in the upper parts of the plant.

These reproductive bumps are associated with the consequences of sexual reproduction. Following fusion of male and female nuclei, the essential sexual event, (see previous page), instead of dividing in the carpogonium, the fusion nucleus in the Gigartinales is first transferred through a fine filamentous outgrowth from the carpogonium to another specialized cell in its vicinity, termed the auxiliary cell. The carpospore-bearing filaments (collectively, the carposporophyte) grow forth from the auxiliary cell in a cluster, and not directly from the carpogonium.

The clusters of spore-bearing filaments produce diploid carpospores which are shed and grow into diploid plants. At maturity, the diploid plants, identical in appearance to the haploid plants, produce spores among their cortical filaments, but each such event is preceded by a reductive division so that these spores (tetraspores) are haploid. They grow into new haploid sexual plants, completing the cycle (see diagram on page 136).

Polyides rotundus X 1.0

Polyides rotundus

Gigartinales

Derivation: "many-formed."
Range: Long Island northward.

This alga is unusual enough to catch your eye as it floats in the wash. Each individual plant of *Polyides* (we have only one species) grows as a small, stiff, bushy mound, two or three inches high. The first axis is affixed to the substrate by a small pad-like attachment and grows straight up in the water. At a quarter to a third of the height of the plant, it branches at a fairly acute angle into two equal axes. Each of these branches again, similarly, until the branching pattern has been repeated six to eight times at increasingly shorter intervals from base to apex. Throughout the entire plant, the axes are round, of nearly equal diameter, and dark red to almost black in color. Each axis is approximately 1 1/16th inch in diameter. Some reproductive plants (diploid) have discrete swellings in the upper branches at the time of reproductive maturity.

Polyides is a deep water alga, and unless you dive for it, you are unlikely to find it except in the wash. Fortunately, its cartilaginous texture is resistant to decay and plants remain in good condition for several days after breaking loose from their attachments. If the specimens we collect in the wash are representative, *Polyides* does not become heavily loaded with epiphytes despite its firm substance.

Even though they display considerable difference in external appearance, the red algae above the level of the first Order (Bangiales) are essentially stereotyped in structure. They are all composed of interwoven filaments. The simplest development of this structure is seen in an organism such as *Nemalion* in which the individual filaments are loosely interwoven and free to move somewhat in relation to one another. *Polyides* has much the same basic construction, but here the constituent filaments are tightly packed together and can only be torn apart. As in *Nemalion,* the plant as viewed under the microscope is divided into a medulla of colorless filaments running mainly lengthwise of the axis, clothed by a cortex of branched and rebranched outwardly-directed filaments arising from them and oriented radially about the axis. The cells of the latter are well provided with chloroplasts.

Cystoclonium purpureum X 0.45

Cystoclonium purpureum Gigartinales

Derivation: "bladdered branches."
Range: New Jersey northward.

This species and the variety described below are each among the commonest of the bushy red seaweeds in our waters. It may well be the first red alga you will collect. Plants are up to two feet long, copiously branched and rebranched, and conspicuous, both where they grow attached near low tide, and in the wash at the water's edge. Growing plants are usually deep red, brown-red, or almost black. Scenescent plants may bleach. Mature specimens may bear conspicuous lumps, the fruiting structures, in the branches. These resist bleaching. Branches are inserted irregularly and often thickly on the axis bearing them. In diameter, branches are mostly between one tenth to the same size as the axis bearing them. The largest axes are not more than a quarter inch in diameter. The smaller branches taper at either end.

A particular variant, often more common than the typical plant, has been given the name *Cystoclonium purpureum* var. *cirrhosum*. In this variety, the ends of some or many of the ultimate branches are elongated and twisted like a pig's tail or the tendril of a grape vine. The identity of the variety is beyond doubt when you see one of these, since this is the only seaweed of any kind to show this distinctive phenomenon. Some old plants without the "pig's tails" may have had them earlier. In other words, some specimens identified as the species itself may really be specimens of the variety instead.

Here are some comparisons that may help to distinguish among the bushy, cylindrical red seaweeds. *Cystoclonium* and *Gloiosiphonia* (p. 153) typically have hundreds to thousands of fine, ultimate branches; the plants look delicate. *Agardhiella* (p. 161) and *Gracilaria* (p. 175) usually have fewer than hundreds of ultimate branches, and the axis may be flattened where major branches arise. *Dumontia* (p. 149) has main branches which are distinctly larger than the main axis, as does *Agardhiella,* but has fewer of them (usually less than ten), and they rarely rebranch as in *Agardhiella.* Branches in *Gracilaria* are not normally larger in diameter than the axis bearing them, though they may occasionally equal it. Branches in *Cystoclonium* and *Gloiosiphonia* rarely approach in diameter that of the axis bearing them; they are cylindrical throughout (but the main axis in *Gloiosiphonia* may become flattened). Each succeeding order of branching is conspicuously smaller in *Gloiosiphonia,* so the main axis of the plant strikes the eye as very much larger than the ultimate branches. The diminution in size of successive orders of branches is gradual in *Cystoclonium* and the extremes of size are less conspicuous, therefore, than in *Gloiosiphonia.*

Agardhiella subulata X 0.60

Agardhiella subulata

Gigartinales

Derivation: honors G. Agardh, a Swedish phycologist.
Range: north to New Hampshire.

Cape Cod marks the effective northern limit for this plant of spring and summer, though occasional specimens may be found in local areas of warm water farther north. It extends in the other direction to the tropics. Distinguishing among the half dozen bushy, cylindrical, branching red seaweeds of six inches to a foot and a half in height, is difficult. When in doubt, the specialist can make microscopic preparations that show the internal structure or reproductive details, from which an accurate determination can be made. The average collector must rely instead on descriptions and illustrations of gross characteristics, or those visible with hand lens.

Agardhiella subulata (our only species) grows in quiet water just below low tide. It is a coarse plant, usually six inches to a foot tall, and deep rosy red in color. Here it is shown attached to an oyster shell. The main axis bears lateral branches to the third order (degree) or more. Branching pattern is mostly alternate, though in some places several branches may all come from one side of the axis. The main axis may be evident throughout much of the height of the plant, or it may "dissolve" into branches. The main branches somewhere along their length usually equal or exceed the main axis in diameter. Most branches taper distinctly toward the point of attachment. When the branches are large in relation to the main axis, the plant looks particularly coarse and top-heavy.

Agardhiella is quite variable. One variant, associated with still, brackish water, is much slenderer than the typical plant, and much more branched. Its branches are less distinctly tapered to the base, and are less conspicuously large in diameter in relation to the diameter of the main axis. You will have to rely on color (deeper red) and, if lucky, fruiting material to separate this from thin forms of *Gracilaria.* You may also find puzzling specimens of what otherwise looks like *Agardhiella,* but are clothed throughout by numerous, short, horizontal branches. This is the diploid plant (tetrasporophyte, see *Gelidium* and *Callophyllis*), in which the spores have not been released but instead have germinated in place. The short "branches" are actually germlings of the haploid stage (sexual plants).

Fruiting bodies form conspicuous swellings in the branches of mature *Agardhiella* (haploid plants) which, however, project only slightly from the surface. In fruiting *Gracilaria,* the swellings are smaller, but half or more of each projects beyond the surface of the axis bearing it. This characteristic will distinguish between these two plants with certainty if you are lucky enough to find fruiting material of sexual plants. A number of other useful comparisons are given it the discussions of *Cystoclonium* and *Gracilaria.*

Hypnea musciformis X 0.60

Hypnea musciformis

Gigartinales

Derivation: after *Hypnum,* a moss
Range: East Coast north to Cape Cod

Hypnea, with only one local species, is another of the relatively large, coarse, bushy, cylindrically-branched red seaweeds, but it has three characteristics which set it off from all others. First is its general appearance, especially the color. Plants grow below low tide in warm quiet water. They have few to many axes arising mostly near the base of the plant. Each axis is fleshy and fragile, something like the texture of pickled watermelon rind, and yellow, or yellow with a purple cast. Secondly, the branches, especially in the outer parts, bear numerous, short, almost thorn-like lateral branches scattered over their surface. These tend to be darker in color than the axis bearing them. Finally, the tips of some of the main branches are naked, swollen, and curved into a hook. These hooks are similar to those on *Bonnemaisonia,* but otherwise the two seaweeds are quite different. Each sharp-pointed hook looks like the hooks used by farmers in handling bales of hay. They are not always plentiful, but usually one or more can be found to confirm the identification.

Cape Cod represents the main northern limit for *Hypnea* as it does for *Gelidium, Agardhiella, Gracilaria, Sargassum, Bryopsis,* and others. *Hypnea* extends southward into the tropics. It is easy to see how the first persons to study East Coast seaweeds conceived of Cape Cod as a dividing line between northern and southern floras. Just as the plants named above rarely extend north of the Cape, such seaweeds as *Fucus distichus, Alaria esculenta, Laminaria longicruris,* and *Plumaria plumosa* are rarely if ever seen south of the Cape, though they are common to the north. The Gulf Stream just to the south, and the comparatively shallow waters of Nantucket Sound, Vineyard Sound, and Buzzards Bay which warm readily in summer, contrast with the cold water along the northern coast and north shore of the Cape. Temperature appears to be the major factor in determining the distribution of plants which relate to these characteristics, geographically or seasonally. Distribution of the large number of seaweeds which are found in quantity on either side of the Cape, such as *Hildenbrandia, Corallina, Pilayella, Scytosiphon,* and many others, obviously is not controlled primarily by temperature but by other factors.

In recent decades it has become clear that isolated areas of warm water occur north of Cape Cod, and of cold water to the south. The entire north shore of Prince Edward Island, for example, is bathed in water that reaches 70 degrees Fahrenheit or more during much of the summer, and there you will find many of the algae familiar south of Cape Cod. As already mentioned, Gay Head is near a finger of cold water. Here and at Montauk on Long Island, you are apt to find algae that "belong" further north. These geographic "exceptions" reinforce the idea of the importance of temperature in determining the distribution of many seaweeds.

Phyllophora pseudoceranoides *Phyllophora truncata* X 1.0

Phyllophora pseudoceranoides Gigartinales

Derivation: "leaf-bearer."
Range: Delaware northward.

We have two species of *Phyllophora,* the one named above, and *P. truncata.* Both are stalked in the lower parts and have expanded blades in the upper parts. Both attach to rocks and other substrates below low tide to a considerable depth. Both form extensive populations and are found throughout the year. Both are used extensively for attachment by other plant and animal organisms. Together, they are quite distinctive and are unlikely to be confused with any red seaweed other than *Chondrus* (Irish moss), but it is often difficult to distinguish one species of *Phyllophora* from the other. Both species of *Phyllophora* have axes which are stem-like and round in the lower part of the plant. In *Chondrus,* the blade tapers slowly to the base and there is little or no portion which is exactly round in section.

The blades of *Phyllophora* are not particularly delicate. They have considerable substance and are dark red to almost black in color if not bleached. An oft-cited difference between the species is the relation of stalk to blade. *Phyllophora pseudoceranoides* tends to be a larger plant, to a half foot tall, and the blades taper slowly into the stalk. The plants of *P. truncata* are rarely more than four inches tall and the taper between stalk and blade is usually more abrupt. Typical plants are shown in this illustration, *P. pseudoceranoides* to the left, and *P. truncata* to the right. Unfortunately, the chances of finding plants which look exactly like these is not great. Variation occurs in the abruptness of taper in either species and an extreme specimen of one species may be most like the description of the other in this regard, and vice versa. Specimens vary also in the amount of proliferation they display, if any. Proliferations are the small subsidiary blades or projections which are often found growing from the tips or edges of mature blades. They are shown well in this illustration of the plant of *Phyllophora pseudoceranoides.* Just as the taper at the base of the blade varies greatly among specimens, so does the taper at the outer edge of the blade. Shown here, both plants have quite wedge-shaped tips, with blunt ends. In some cases, however, the outer tip attenuates in width in similar fashion as it narrows at the point of attachment. The only sure way of distinguishing between the species is by means of reproductive detail.

No one knows why or how these two species show so much variation. It would be worthwhile to examine the relative influence of genetics and environment in determining the form a given specimen of either of these species takes.

Mastocarpus stellatus X 1.0

Mastocarpus stellatus Gigartinales

Derivation: "bumpy bodied."
Range: North Carolina northward.

Mastocarpus stellatus is not as common as *Chondrus* (see next), but where it is found, it is often next to, or mixed in with a population of *Chondrus*, and the two may be quite similar in general appearance and color. We have never seen any healthy, but sun-bleached *Mastocarpus*, though this phenomenon is relatively common in *Chondrus*. One characteristic, that of the fruiting structure, separates the two with certainty. In *Chondrus*, plants either show no obvious modification for fruiting or localized areas of thickening ("bumps") within the blade. In *Mastocarpus*, on the other hand, many plants in any population will have surfaces covered with tiny papillate outgrowths. The reproductive organs are localized in these nubs. Another character to separate them is that the blade edges of *Mastocarpus* roll in somewhat, while *Chondrus* has flat edges.

Still another useful characteristic for distinguishing a population of *Mastocarpus* from *Chondrus* is iridescence. On a bright sunny day, unbleached plants of *Chondrus* usually display a shimmering, glancing purple color from shifting spots on the fronds as long as they are underwater. This is caused by a film on the surface of the plant which in sunlight acts like colored mirrors. No one knows if iridescence is of any real value to *Chondrus*, but no other East Coast algae possess this characteristic.

Mastocarpus stellatus was known for a long time as *Gigartina stellata*, the only representative of this genus found in the North Atlantic but a genus with many representatives on the Pacific coast. The change of its name was the result of studies by John West and his students on the life cycles of *Gigartina* species. While some of the species have haploid (gametophyte) and independent diploid (tetrasporophyte) phases with similar forms as is characteristic for *Gigartina*, the investigators found that other species had an encrusting diploid phase and only the haploid phase formed a blade. These species, in which the phases had distinctly different forms, were reclassified in the genus *Mastocarpus* and thus our *Gigartina stellata* became *Mastocarpus stellatus*. The encrusting phase of *Mastocarpus*, previously called *Petrocelis middendorffii* (in a situation similar to what happened with the *Trailliella* phase of *Bonnemaisonia*), is a very dark leathery crust that grows around bases of *Chondrus* and the haploid phase of *Mastocarpus*.

The story of *Mastocarpus stellatus* in our area has still another twist. It appears that it often "short circuits" its life cycle. Female haploid plants develop carposporophytes and produce spores (carpospores) without the need for fertilization of their carpogonia by spermatia (male gametes). In the absence of sexual fusion, the carpospores are haploid and develop into more haploid blades rather than into *Petrocelis* crusts. In fact, in some populations of *Mastocarpus*, male haploid plants are completely lacking.

Chondrus crispus X 0.50

Chondrus crispus

Gigartinales

Derivation: "cartilaginous."
Range: New Jersey northward.

When the first author of this book was a child, children were expected to consume gallons of milk. Milk in moderation is a fine food, but some children don't like it. One of the ways New England mothers attempted to get large quantities of milk into their children was by disguising it as Irish moss blancmange. This is a bland milk gel which can be sweetened and flavored in various ways. Kingsbury hated it.

As the second author of this book can testify, Kingsbury has inflicted Irish moss blancmange on hundreds of students in his classes since then. His wife thinks he is working out his early frustrations. Many students seem to like it, however, and the demonstration of how much milk can be gelled with a little Irish moss is an impressive presentation of a characteristic of the basic raw material which now stands behind a multimillion dollar industry.

If you would like to try New England blancmange, rinse wet (fresh) Irish moss, or soak dried moss in fresh water until soft. Discard any that is not reasonably clean and free of attached animals. Throw a small handful (a generous half-cup) into a quart of milk in a double boiler and cook for 25 minutes. Strain the moss out with cheesecloth. Stir in some salt, sugar, and vanilla to taste. Pour into molds; chill until firm. Serve as is or with sugar and cream.

Be sure you have Irish moss. Most others won't work. *Chondrus* (only the one species) is not difficult to recognize, though it varies tremendously in general appearance as shown in the illustrations. Individual blades are two to six inches tall, flat in a single plane (or gently twisted) narrowing downward to the point of attachment, but with little or no round stalk at all. The fronds are of cartilaginous substance, appreciably thick, and repeatedly branched in a pattern of "Y"s. Populations vary in the degree to which the branches are "pulled out", in the width of the branches, and in color. Some plants, particularly to the south of Cape Cod, may have very long, narrow branches, narrowly branching. Others, particularly farther north, have wide blades, widely branching. Plants growing at some depth or in the shade of rocks are deep red, almost black. Plants that grow fully exposed to sun between tides or just below low tide may be yellow-green. These may look dead, but may be actively growing.

In the life cycle of *Chondrus*, the haploid phase (gametophytes) and the diploid phase (tetrasporophytes) are of similar form, but they differ slightly in their ecology. Diploid plants tend to grow in deeper water than the haploid gametophytes. *Chondrus* competes well as shown by its abundance (sometimes covering acres of stony bottom) and the range of habitats where it grows. Once established, a population of *Chondrus* is perennial, growing most in summer but persisting through the winter, and a healthy population keeps out potential invaders effectively by occupying all available space. Perhaps because of its tough construction, *Chondrus* is not eaten much by the common periwinkle (*Littorina littorea*), while potential invaders such as *Ulva* and *Enteromorpha* are favorites.

Vegetable gums are used in large amounts in such industries as foods, textiles, paint, paper, pharmaceutics, and cosmetics. The gums (technically, phycocolloids or hydrocolloids) obtained from certain seaweeds are indispensable for certain purposes. Principal gums from seaweeds include agar (a Malay word meaning "jelly") obtained primarily from *Gracilaria* and *Gelidium*, algin from the brown alga *Macrocystis* and to a limited extent from *Ascophyllum*, and carrageenan primarily from *Chondrus* and *Eucheuma*. Of these, most agar is harvested in Asia and Latin America and processed in the Orient. Algin is produced commercially in large amount on our West Coast. Carrageenan comes primarily from *Chondrus* in the North Atlantic and from *Eucheuma* grown in the Philippines.

Kingsbury remembers seeing brewers scatter raw Irish moss on the surface of the brewing vats at the Boston Brewery years ago. It formed a floc (gel) which settled to the bottom, carrying impurities with it. A generation ago crude Irish moss extract was used in hand lotions and was considered beneficial for chapped hands because of its ability to soften the skin. These crude uses are a long way from the refined extractives now available commercially.

These algal gums are constituents of cells walls. Mixed with water, they form viscous solutions or jellies. Starting with the raw material extracted from seaweeds, industrial chemists are now able to manipulate solubility and other characteristics and produce hundreds of consistent high quality products. They are used in such things as toothpaste, shaving creams, hand lotions, pie fillings, custards, ice cream, salad dressing, and chocolate milk; and as bulking or slow release agents for drugs, as textile and paper sizing, paint thickeners, and in literally hundreds of other ways. The human digestive system lacks enzymes able to digest these phycocolloids so they have found use as a bulking agent in low-calorie foods.

At the same time that much effort has been spent on improving techniques of extraction and the characteristics of the extractives, far less work has been done on improving methods of husbandry or of harvest. In many cases, naturally growing populations of seaweeds are still harvested by hand or with the aid of rakes. This contrasts with the sophisticated mechanical harvesters used to collect *Macrocystis* (for algin) on the West Coast. Most of the carrageenan today comes from *Eucheuma* in the Philippines. Here in a labor-intensive cottage industry, *Eucheuma* is farmed by tying plants to monofilament lines strung in warm, clear water where there is adequate water movement but not too much wave action. After two to three months, plants are harvested by hand.

Details of the processing procedure for carageenan are company secrets, but in general, harvested seaweeds are dried, often just by spreading on the ground in the sun, and then baled for shipment to processing plants. There the algal material is digested and the carrageenan extracted using hot, alkaline water, and concentrated in steps. Different extracts varying in their gelling properties and viscosity are blended for the final product. Interestingly, *Chondrus* has two slightly different forms of carageenan with different properties. One form comes from the haploid generation and the other from the diploid generation.

Chondrus crispus *variation of form of frond* X 1.0

Ahnfeltia plicata X 1.0

Ahnfeltia plicata

Ahnfeltiales

Derivation: honors N. I. Ahnfeldt of Lund, Sweden.
Range: New Jersey northward.

Of all red algae, this is the stiffest. It grows at some depth below low tide around Cape Cod and is most likely to be collected in the wash where it comes ashore in small bundles a few inches long. These discrete clusters resemble the bundles of faggots collected by peasants in the forests of Europe at an earlier time and carried home on donkey back. Each bundle is composed of a tangle of plants. The individual plants are much branched, cylindrical throughout, dark red or almost black unless bleached, and very stiff considering the diameter of the axis which is less than 1/16th inch.

Ahnfeltia grows in groups in rock crevices and other protected locations. In Cape Cod waters the plant lives at some depth, but farther north it can be found in some of the lower tide pools. It is resistant to decay, although clusters bleach white after a while in the wash. Tangles of bleached *Ahnfeltia* may be confused with similar tangles composed of the roots of *Spartina* (salt marsh grass) which sometimes wash out of the turf at the edges of streams or ditches in salt marshes and are carried out to sea, later to appear in the flotsam at the edge of an incoming tide. *Spartina* roots, even well-tumbled clumps, usually have a number of smaller rootlet-branches by which they can be recognized. *Ahnfeltia* tangles are about the same diameter throughout.

The life cycle of *Ahnfeltia* is similar to *Mastocarpus*. Plants with stiff, wiry branches are the haploid phase (either male or female gametophytes), and as a result of the sexual process and release of carpospores, the alternate diploid phase is produced, which is an encrusting form (formerly called *Porphyrodiscus*).

In recent years, use of the electron microscope has allowed examining fundamental cell structure in great detail. This, with an increased ability to culture red algae through their life cycles and study reproductive detail (see *Callophyllis*), has lead to a better understanding of red algal classification. In red algae (with only a few exceptions) adjacent cells in a filament are connected by a pore normally blocked by a large plug. Viewed with the electron microscope, the plug has a girdling groove that "locks" it into the pore. Dr. Curt M. Pueschel and his colleagues have examined the pit plugs of a wide range of red algae. Features associated with these pit plugs, especially the number of outer layers (called "caps") and the presence of a membrane, provide stable characteristics for defining Orders of red algae. Members of a given order will show a consistent combination of these plug features.

The pit plug of *Ahnfeltia* doesn't match those of the Order in which it was formerly placed (Gigartinales). Hence *Ahnfeltia* is now placed in a separate Order of its own, the Ahnfeltiales, on the basis of its unique plug characteristics and reproductive cycle. Despite difficulties of preparation and technique in using an electron microscope, this new window on red algal basic structure is forcing a reevaluation of fundamental relationships.

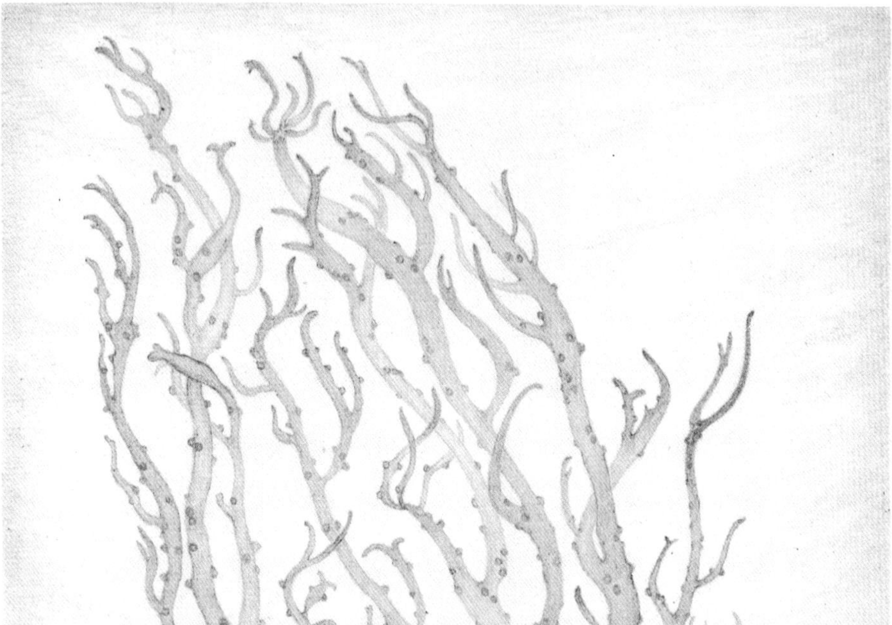

Gracilaria tikvahiae X 0.60

Gracilaria tikvahiae

Gracilariales

Derivation: "graceful."
Range: Canadian Maritimes south.

This is the only species of *Gracilaria* found in our area. It is a coarse, bushy seaweed, which may be a foot or so tall, and is highly variable in general appearance. Its branches may be flattened or round in cross section, and its color ranges from dull purple to yellow-greenish. A typical specimen is shown here. The lumps or swellings, resulting from the development of the dependent diploid phase (carposporophyte) following fertilization, identify it as a haploid female plant. See further comments under *Agardhiella* and *Cystoclonium*, with which *Gracilaria* may be confused.

Ideally, the name of a species should not change, but must sometimes be altered as more is learned about it. This has been especially true for red algae as a comparison of the first and second editions of this text will readily show. *Gracilaria tikvahiae* is a good example of why such changes are made.

In the second edition of William Randolph Taylor's monograph on the marine algae of the northeastern coast of North America, published in 1957, *Gracilaria* was divided into two species, *G. verrucosa* and *G. foliifera*, based on similarities to these species in Europe. Subsequent investigations of American algae showed that ours was actually only one highly variable species. Further investigation into its life history (development of the carposporophyte) revealed that our American species was fundamentally different from either of the European species despite a similar appearance. In 1979, Jack McLachlan proposed the present name *Gracilaria tikvahiae* for the *Gracilaria* found in New England and the Canadian Maritimes. He chose the species name to honor Dr. Tikvah Edelstein, a researcher in his laboratory, whose premature death ended a promising career in the study of red algae.

Recently, the validity of the separation of the European *G. verrucosa* from the North American *G. tikvahiae* has been tested by molecular techniques, in which DNA was extracted and specific genes compared. The differences in these genes confirmed the separation of the two species made originally by the more traditional approach of observing reproductive structures.

Lomentaria baileyana X 1.0

Lomentaria baileyana

Rhodymeniales

Derivation: "pod-like with joints."
Range: East Coast north to Canadian Maritimes.

This is a delicate alga of quiet, shallow water, growing at and below low tide. It is primarily a southern form, which rarely grows north of Cape Cod, but it is fairly common in Buzzards Bay and the Sounds. Where conditions are favorable, it forms open mounds, an inch or two tall, on rocks and shells. One or more main axes arise from the rock and are repeatedly branched and rebranched. One of the most distinctive characteristics of the plant is the fact that the branches of all levels tend to be gently and regularly curved. A long branch may describe a complete semicircle; the shorter ones, gentle arcs. In the outer parts of the plant, the ultimate branches tend to come out on the outside of the curve of the axis bearing them, all on the same side. All axes, even the ultimate branches, taper at both ends. The taper toward the free tip is gradual and the tip is sharply pointed. The taper at the other end, toward the point of attachment, is usually gradual but in some forms may be quite abrupt. The plant varies quite a bit, and some collections may not have particularly curved branches.

Each branch is hollow except at the point where it joins the axis bearing it. In structure, the axis begins like a *Nemalion* with several filaments running lengthwise and giving off filament-branches which collectively make up the cortex. In *Lomentaria,* however, a cavity develops early among the original central filaments in a regular predetermined manner, not like the tearing apart of others as in *Gloiosiphonia.* The cortex, also, is much modified in comparison with that of an alga like *Nemalion,* and its derivation from a fundamental filamentous condition is not immediately obvious under the microscope.

Lomentaria has a distinct tendency to lighten in color as it dries. Mounds of it as collected are dull red or dark red. Dried specimens seem much redder. On drying, it sticks well to the paper and its gentle curves can be put to good use in creating satisfying designs on note paper if you are inclined to decorate messages thus. A lucky collector may find a specimen of *Lomentaria orchadensis.* This rare seaweed is basically similar to the former except that the branches are fatter and arranged in two rows (pinnate) on the main axis.

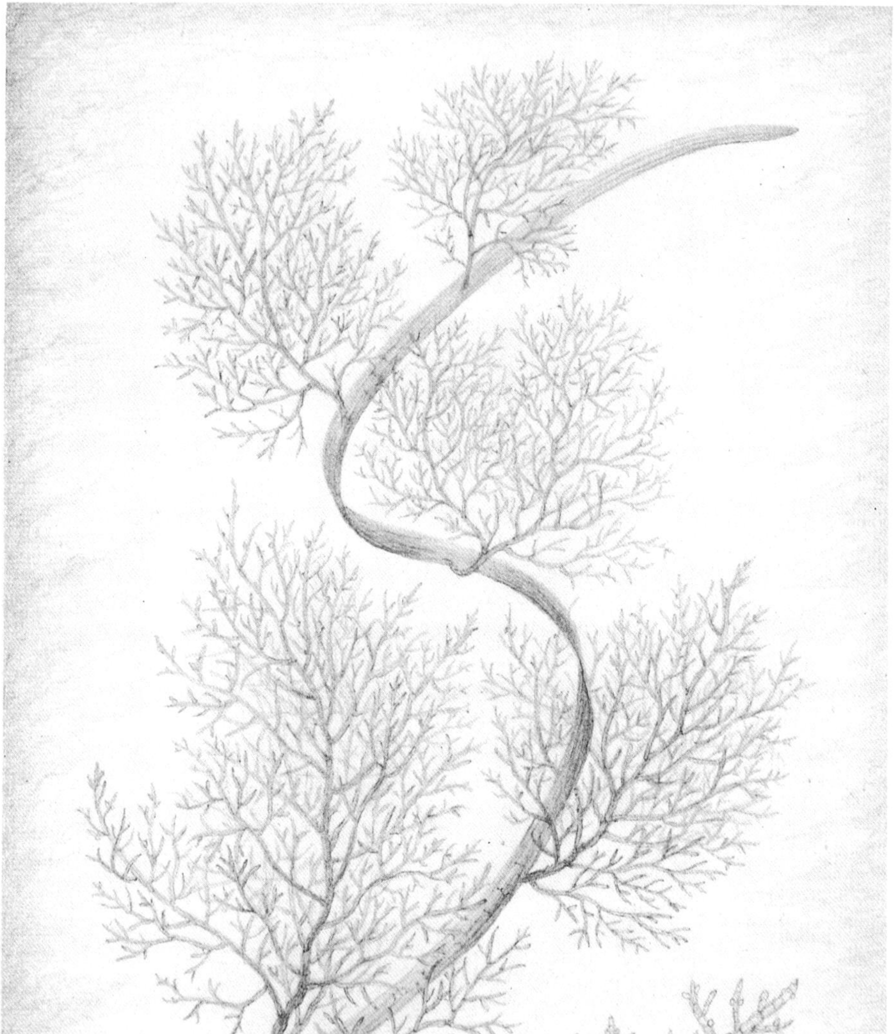

Champia parvula X1.0 on *Zostera* enlarged detail

Champia parvula

Rhodymeniales

Derivation: honors M. Deschamps, a French botanist.
Range: East Coast north to Canadian Maritimes.

Champia is a very common midsummer seaweed along the south side of Cape Cod and at the Islands. It is another of that group for which Cape Cod represents the effective northern boundary, although occasionally *Champia* has been reported as far north as Canada. The plant is densely branched, forming intricately tangled tuft-like growths one to three inches tall attached to stones, shells, eelgrass, and coarser seaweed in quiet water below low tide line. In this illustration the tufts have been pulled apart leaving individual, opened-out plants. Plants may be dull red, brownish, pale pink, or greenish. When removed from the water the turgid, water-filled stalks remain more or less rigid. Axes of all orders are hollow, but crossed by diaphragms of tissue at regular intervals. The compartmentilization so effected is usually marked on the outside of the plant by a constriction around the axis at the level of each diaphragm. The cylindrical segments thus delineated may be of equal height and breadth or one and a half times taller than wide. They are shown in the magnified insert.

Branches arise at the constrictions and narrow toward their tips, but the relative proportions of the segments of which they are composed remain constant. You will need a hand lens to follow this out to the very tip. Under some growing conditions the plants are not particularly constricted at the diaphragms, but the segmented nature of the axis can usually be seen in such plants by holding them up to the light.

Bays of the sort in which *Champia* grows may have quite different characteristics from more open waters. One of the most important factors necessary to the growth of any photosynthetic organism is light. Light under water is significantly changed from light as it reaches the surface of land. Passage through marine water not only reduces the intensity of light, but also modifies its color. Most of both effects can be traced to the presence in seawater of particulate matter. Particles range in size from the tiniest accumulations of matter to the microscopic living organisms of the plankton. Some of the smaller particles may be mineral, but the great majority are either living or the product of living organisms. They often are pigmented or have pigments in them and therefore absorb the colors of white light differentially. What is left for the seaweeds at the bottom is what has passed through, passed by, or been reflected by the colored particles in the water.

In the open ocean, two periods of greater fertility per year usually result in a greater abundance of plankton and other particulate matter in seawater. These come roughly in spring and fall. In bays, however, particulate matter commonly increases as the water warms, and may attain greatest density during the summer months. This may be a local phenomenon, obscuring light penetration in one bay, and not in another near it, or even in one arm of a bay as opposed to the rest. From a boat, persons are not usually aware of the turbidity of the water under them. Try diving in such water. Sometimes it is so dense that your hand goes out of view if you extend your arm in front of you.

Filamentous species

The Ceramiales is considered the most advanced of Orders of red seaweeds, but among its members are seaweeds whose structure is as simple as any. These are the filamentous forms, whose axes are composed of single chains of barrel-like cells placed end to end. They are considered advanced on the basis of certain features of reproduction. Simplicity is not always correlated with primitiveness but is often the result of evolutionary reduction from a more complex ancestry. In such evolutionary series, whether in plants or animals, the reproductive mechanism is often the slowest to change, and therefore, the most fundamental characteristic from which to interpret change. In possessing filamentous structure, these forms have come full circle from the primitive filamentous members of the red algae, exemplified in this book by *Audouinella purpurea,* but their reproduction is far more complex. The first task in recognizing these forms is to separate them as a group from the other red seaweeds. The second task is to separate and recognize the individual genera among them. Neither of these can be accomplished with absolute certainty except by using a microscope, but some generalizations will help in approaching these goals.

The primitive filamentous red algae, exemplified by *Audouinella*, are mostly tiny to very tiny, forming small mounds or tufts of red threads rarely more than a half inch high. The filamentous seaweeds of the Ceramiales are mostly larger (though not always). The conspicuous, branching, filamentous red algae which are common in the summer almost always belong here. However, many of the seaweeds that look filamentous at first glance, such as *Spyridia, Polysiphonia,* and delicate species of *Ceramium,* have very fine but nonetheless multicellular axes, and therefore are not strictly filamentous. Look closely with a hand lens at the largest axis you can find in the middle (not bottom) of the plant. If its surface looks lacy, particularly at branch points, or has any hint of longitudinal lines along the surface, the plant is not truly filamentous (not discussed in this section). Depending on size of cell and power of hand lens, you may or may not be able to see the individual cells of a filament. If the threads are very fine and you cannot see individual cells, the filaments are probably too small for the cells to be visible. If you cannot see individual cells in a coarse or heavy "filament", be suspicious that it is a multicellular axis composed of very tiny cells. In many species, a plant with a multicellular axis toward the base becomes filamentous at the branch tips.

Filamentous and sheetlike seaweeds normally grow much more rapidly than coarser algae. In filaments or sheets all the cells are fully exposed to the light and photosynthesize. In contrast, photosynthetic activity in coarser algae is confined to an outer layer (cortex) which must provide nutrition to support the nonphotosynthetic tissue in the center (medulla). However, there are tradeoffs. Filaments and sheets, with their relatively simple construction, are more susceptible to physical damage and to predators, while the thicker, structurally complex algae are more resistant to these stresses. Filaments and sheets, however, are able to compensate for the loss of tissues by growing back rapidly.

The following discussion characterizes the most common species for each of the common filamentous genera of the Ceramiales found in Cape Cod waters and shown in the illustrations.

Callithamnion corymbosum

Ceramiales

Derivation: "beautiful bushlet."
Range: entire East Coast.

This alga forms very delicate spherical masses that, in the water, are like clouds of pink, or pink cobwebs. It grows on eelgrass and on various algae below low water to some depth. The individual masses are one to three inches in diameter. They may become detached from their hosts and float in on the wash. The plant grows in early summer. As it ages in mid summer or later, it often loses its obvious red color, becoming dusty red or brownish, and eventually bleaching more or less. To determine the branching pattern, spread a tuft out in shallow water. Branching is irregular to alternate among the larger axes. The smaller ones toward the periphery of the plant usually divide equally into two. The outermost, very fine branchlets tend to be grouped in broom-like clusters. This is shown clearly in the illustration.

Plenosporium borreri (derivation: "many spored;" range: Delaware to Massachusetts) is separated from *Callithamnion* by the number of spores per cluster (more than four) in the diploid phase in the life cycle. Otherwise, this plant (not illustrated here) is nearly indistinguishable from species of *Callithamnion*. Plants are mostly one to two inches tall, with dense to open branching, and grow attached to various algae, pilings, or rocks at or below low tide. When spread out in a little water, a typical plant displays a broadly triangular pinnate cluster of branchlets at the tip of each major branch which is mostly naked beneath.

Callithamnion corymbosum X 1.0 on *Zostera* — *enlarged detail*

Spermothamnion repens — Ceramiales

Derivation: "seed-bearing bushlet."
Range: Virginia to Nova Scotia.

The "seed-bearing" designation is an anachronism dating back to the time before reproduction was understood in the "simple" seaweeds. Linnaeus, the organizer of our present classificational scheme, lumped the algae with a number of other organisms as "cryptogamia" because they didn't have flowers. The word means "hidden reproductive parts," or more accurately, "hidden gametes." It has turned out of course that in no group of plants are the gametes less hidden than in the algae. *Spermothamnion* was separated originally from *Callithamnion* because its spores are borne separately and obviously. In *Callithamnion*, they are massed together in a gelatinous aggregation.

Spermothamnion repens will catch your eye. Each plant is a dense spherical tuft of deep red filaments, about the size of a golfball. It is common in the waters to the south of the Cape where it grows from low tide to deep water, attached to *Chondrus* or *Phyllophora*, often coming free. The free-floating masses persist for some time in the water and often accumulate in large numbers at the edge of an incoming tide. With hand lens, you can determine that the branching pattern is different from that of *Callithamnion*. In *Spermothamnion* the lateral filaments come out oppositely from the axis bearing them, in pinnate fashion.

Spermothamnion repens X 1.0 on *Phyllophora* — enlarged detail

Antithamnion cruciatum

Ceramiales

Derivation: "opposite bushlet."
Range: entire East Coast.

Antithamnion is generally similar in habit to the algae above. It is separated from them by the possession on the filaments of gland cells. These are tiny, colorless cells that develop from place to place from the lateral surface of a filament. They are so different from the rest of the cells of the plant that they were once thought to be parasitic fungi living on the alga. Unfortunately, they are too small to be seen without the aid of a microscope. The related genus *Antithamnionella* lacks gland cells.

Antithamnion cruciatum is less branched and forms more open growths than either of the two plants discussed first. The tufts are one or two inches high and grow on rocks and coarse seaweeds at and below low tide line. The most distinctive feature of this plant is the fact that the major axes bear short branches in pairs, each member opposite the other, or in fours, each at right angles to the other three. The terminal filaments are very delicate and very clustered so that, spread out in a little water, the plant gives the appearance of having spots of color toward the edges.

Antithamnion cruciatum branch greatly enlarged

Seirospora interrupta Ceramiales

Derivation: "chain of spores."
Range: New Jersey northward.

This is a large but soft alga. The technical distinction separating it from the others is that it produces its spores in open clusters of linear chains. The plant (with only one species) is two to six inches tall and broadly triangular in outline. A major axis is usually discernible from base to tip. The coarsest branches are often corticated with small, superficial cells that obscure the basic filamentous nature of the plant. The major lateral axes are alternate and the terminal branchlets somewhat clustered. The plant is quite common south of Cape Cod. It grows on eelgrass, coarse algae, shells, and stones below low tide.

Seirospora interrupta X 10 *enlarged detail*

On light, drying, and temperature as environmental hazards

Red algae may be difficult to recognize simply by their color and sometimes getting the right identification is done best by a process of elimination. The green algae are usually an obvious green color, and many brown algae are distinctly tan (*Fucus* and *Ascophyllum,* however, are common exceptions). Thus if a seaweed is not clearly tan or green, it is likely to be a red alga and may show a purple or dark red coloration near its base if examined closely. The color of shallow-water red algae is particularly variable, as exemplified by two common species *Chondrus crispus* and *Mastocarpus stellatus. Mastocarpus* is often a dark brown color, and *Chondrus* may be dark with "bleached" or light yellow tips. Both of these algae are exposed to high levels of sunlight when they grow in the intertidal region. Under these conditions, obtaining sufficient light to carry out the photosynthetic process is not a problem. Rather there is a greater danger of damage from too much light. To protect themselves, the algae produce their own versions of sunscreens, often in the form of yellow or brown carotenoid pigments. Instead of collecting light for photosynthesis, these pigments act as filters and dissipate excess solar energy. On the other hand, red algae growing in the shade of other algae or in deep water have less need for protective pigments and are more likely to have an obvious red color (see *Phycodrys*).

Exposure to high levels of light is only one danger faced by red algae that spend part of the time out of the water in intertidal habitats. Another is the danger of water loss by evaporation while exposed to the air. Intertidal red algae, such as *Chondrus* and *Mastocapus*, have adapted less well than the rockweeds *Fucus* and *Ascophyllum* (brown algae) to prolonged exposure during low tide and do not extend as far up the shore unless they are in very damp places. One such place is found under the fronds of the rockweeds. If you turn back the fronds, you may find an understory growth of red algae, including crusts such as *Hildenbrandia* and *Phymatolithon*.

Seaweeds that spend part of the time out of water must also tolerate greater temperature variation than in the seawater. This can lead to stress from high air temperatures in summer and temperatures below freezing in winter. For his doctoral dissertation at the University of Maine, Steven Dudgeon studied the relationship between *Mastocarpus* and *Chondrus* in the low intertidal region and found that *Mastocarpus* tolerates low winter temperatures and freezing conditions better than *Chondrus.*

Griffithsia globulifera X 1.0

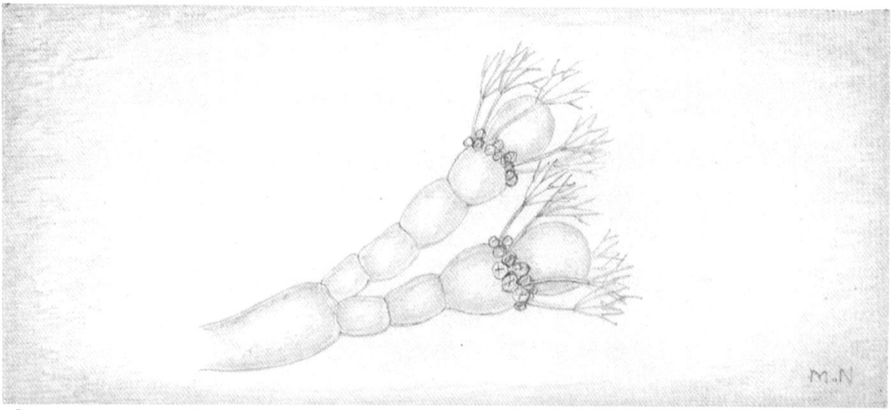

Griffithsia globulifera with tetraspores *enlarged detail*

Griffithsia globulifera Ceramiales

Derivation: honors Mrs. Griffiths, an English phycologist.
Range: East Coast north to Canadian Maritimes.

Although the branches of this alga are composed of a single series of cells, and are therefore filamentous, you will have no trouble separating it from the other filamentous species previously described. Each cell is tremendous, easily visible to the naked eye, and looks like a huge pink balloon under the hand lens. Cells may be spherical or markedly egg-shaped, large end up. They are linked end to end in branching axes. Branches arise from the upper ends of cells.

Each plant is one to two inches tall or a little more, and spreads broadly upward from its basal attachment. Each cell has thousands of nuclei (unusual for red algae) and many chloroplasts. Even with many chloroplasts, each cell is so large that the chloroplasts are rather widely dispersed, and the plant as a whole is delicate pink. The cells toward branch tips bear colorless bristles of thin branching filaments which are soon shed. *Griffithsia* grows on solid substrates such as rock and shells below the low tide line. It is especially common on the south side of Cape God in mid summer, coming in with the wash as inconspicuous, nearly transparent turgid clumps.

Griffithsia is a useful alga to employ in demonstrating reproductive details for the Order Ceramiales. Its reproductive organs are mostly out in the open, while they tend to be hidden or covered with sterile tissue in most other members of the Order. Furthermore, the fruiting plants are sufficiently different that, with practice, one can distinguish among the male plants, the female plants, and the diploid (tetrasporophyte) plants by naked eye.

In male plants, the clusters of tiny, colorless spermatia (naked male cells) are produced on the upper third of each balloon-like vegetative cell near the branch tip. They are densely packed together, and the cell looks like a head with a crew-cut haircut. The female branching system is produced at the joints between cells toward the tips of branches on separate female plants. Each female system consists essentially of a four-celled branch with the carpogonium (female cell) at its tip, together with several sterile, vegetative cells around it. All the cells of this apparatus are much smaller than the balloon-like cells of the axis. The whole female branching system which is twisted in on itself, is in fact smaller than a single cell of the main vegetative axis. The diploid (tetrasporophyte) plants are similar in gross characteristics to the sexual (haploid), but produce orange-like clusters of tetraspores at the points where axis cells join in the outer parts of the plant.

Plumaria plumosa X 1.0

Plumaria plumosa — Ceramiales

Derivation: "feathery."
Range: New Jersey northward.

This seaweed is fairly common north of Cape Cod and is found in occasional cold water locations to the south. It seems to be favored by shaded habitats as on vertical rock faces underneath *Ascophyllum*. A bit of it in the wash will attract attention by its unusual form and branching pattern. The major axes, one to six inches long, are relatively coarse and cylindrical. The rest of the plant is delicate. It is composed of systems of branches which in turn are rebranched, all in a single plane of space, and the general effect is much like the teeth of a comb coming out of both sides. The branches are not only oriented in a single plane, but also regular in placement, giving the frond a feather-like or diminutive fern-like appearance. If you try to analyze the branching pattern with hand lens, you will find it regular, but alternately unequal.

The regularity of structure in *Plumaria* is particularly interesting. This plant is basically filamentous. The thickness of the major axes is produced secondarily by overgrowth of the original filament with smaller randomly oriented cells and filaments which bud out of the cells of the original filament. Branching is frequent and is brought about when a major lateral filament is initiated from a cell of the axis bearing it. Branches normlly appear in pairs, as in *Antithamnion*, but are not initiated at exactly the same time. The one appearing first somehow dominates the other which never becomes as large. In fact, in the outermost portions of the plant, where all the lateral branches are small, the smaller of each opposite pair may be repressed entirely. Furthermore, looking at successive branch-pairs, the larger branch of each opposite pair alternates from one side of the axis to the other. What biochemical or physiological mechanism controls the development of these plants in such a delicate manner as to account for the pinnate regularity and equality, yet at the same time the alternately regular asymmetry, is unknown.

The fruiting structures in *Plumaria* appear grossly as small "bumps" or dots in the upper parts of a plant. The details of sexuality are similar to those described for *Griffithsia* (see previous page). The distinction which separates members of the Ceramiales from the other Orders discussed earlier is the fact that an auxiliary cell is not produced in the female apparatus until after the carpogonium has received the male nucleus and it has fused with the female nucleus. The auxiliary cell, to which the fusion nucleus is transferred, buds forth from the lower part of the female apparatus and develops after fertilization. Short, spore-bearing filaments of small cells are produced from it and the diploid spores shed from these grow into the diploid phase (tetrasporophyte) of the life cycle. A reduction division takes place in ultimate formation of tetraspores, and they give rise in turn to the haploid, sexual generation, completing the cycle.

Spyridia filamentosa X 1.0

Spyridia filamentosa — Ceramiales

Derivation: "basket-like."
Range: East Coast north to Nova Scotia.

In many ways this unusual plant represents a combination of characteristics found separately in other members of the Order Ceramiales. Unfortunately, a microscope is required to do it justice. A well-developed plant of *Spyridia* (only one species) appears filamentous but is unusually large (to a foot tall) and coarse. Its main axes are particularly thick. On removal from the water the plant is stiff and brittle. Actually, it is truly filamentous only in the extremities. Each main axis is built from a filament which is covered secondarily by the growth of cells one layer thick around the central cells of the original filament. Each of the outer cells (pericentrals) is at first as long as the central cell which it covers. Together they look like a layer of cordwood on end and edge to edge, laid vertically around a barrel. Polysiphonous construction is illustrated in its simplest form at *Polysiphonia nigrescens.*

In *Spyridia,* the tiers of pericentral cells become further specialized so that, at maturity, the axis is composed of tiers of short pericentral cells alternating regularly with tiers of long pericentral cells. If your hand lens is good enough to distinguish this characteristic, it serves to identify *Spyridia* beyond question. With less magnification, what looks like an individual cell may actually be a tier of cells.

Young plants, or the younger, outer parts of old plants look as though they need a shave. This appearance comes from the presence of numerous filamentous branchlets all over the surface, each not more than a tenth of an inch long. These filamentous branchlets are lost as the plant ages. Each branchlet is composed of a single file of cells but is encircled by bands of small, spherical cells at each place where two cells join. The band of small cells looks the way a ring made of closely spaced beads might look on a finger.

Spyridia is an alga of warm waters and is common during summer months. Cape Cod represents its effective northern limit, although occasionally it is found farther north. It ranges in the other direction to the tropics. It grows from rocks and shells below low tide line, often coming loose and drifting about on the bottom as it gradually turns straw colored and dies.

Ceramium rubrum on *Chorda* X 1.0

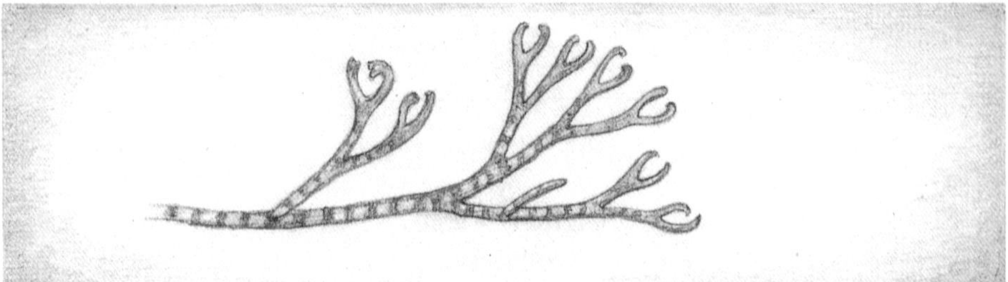

Ceramium rubrum enlarged branch detail

Ceramium rubrum Ceramiales

Derivation: "like a small pitcher."
Range: entire East Coast.

Ceramium rubrum is perhaps the commonest *Ceramium* in Cape Cod waters and in turn is among the most common of all the seaweeds on our shores. Professor Harold Bold, when he took the Woods Hole students on field trips, was apt to call this one "Anyface" after a character in the Dick Tracy comic strip of years ago. Like "Anyface," *Ceramium rubrum* appears in many guises. You may pick it up time after time, even in the same location on the same trip, each time thinking you have something different, only to have it fool you. It comes in almost any color a red alga can assume, from deep wine red, to pink, to brown, to dusty, to yellowish, to greenish, and finally to bleached white. You may find this entire range among plants in the same wash, and it is this variation in color more than anything else which fools you into picking it up again and again. The plant also may vary quite a bit in general habit. It may be in small pieces, or it may occur in large tangles. It may be attached to a variety of substrates or it may be floating. It may be intertangled relatively tightly or not at all. Variation in branching pattern may make it appear distinctly coarser or finer than the last piece you picked up.

Yet when looked at closely, *Ceramium rubrum* usually identifies itself quite readily—at least as a *Ceramium*. Species of *Ceramium* have two main identifying characteristics. They are banded, and the branch tips end in tiny incurved divisions which look like lobster claws. When you pick up a coarsely "filamentous" red alga, look to see if it has zebra-like stripes at regular intervals across the axis. Each axis starts out as a single large-celled filament, but numerous small cells are formed almost immediatey at the junctions between the large cells. These create muff-like rings around the filament. In some species the "muffs" do not grow, and naked filament separates them. In others, including *C. rubrum* the "muffs" grow up and down over the surfaces of the filament within and cover the large cells of the filament completely with a cortication of tiny random cells. Even when this happens, the outlines of the large central cells of the original filament remain obvious beneath the cortication. The species with partial cortication are clearly banded, even to the naked eye. The fully corticated species are less clearly banded to the eye but under the hand lens will display transverse markings. This is most obvious in the youngest branches. The "lobster claw" branch tips are very distinctive, and are usually visible to the naked eye. One caution is necessary. You should have actively growing material. The "claws" are sometimes lost when the plant ceases growth. Even in an old plant, it is usually possible to find a few branches that are still growing. *Ceramium* is most likely to be confused with *Polysiphonia* and *Spyridia* if it is not correctly recognized on first try.

Phycodrys rubens X 1.0

Phycodrys rubens Ceramiales

Derivation: "sea oak."
Range: New Jersey northward.

This and the next two are among the most beautiful seaweeds anywhere in the world. Their beauty is apparent even to the most casual seaside observer, and they are therefore among the seaweeds which are most apt to evoke "What is it?" Fortunately the same characteristics which make them esthetically appealing, make each of them easy to identify.

Phycodrys is a seaweed of deep, cool water along the Cape Cod shoreline. A common habitat for it is on the holdfasts of *Laminaria* plants. It also grows on stones and shells. Farther north it may be found in tide pools and in shallow water.

Phycodrys starts growth as a single, simple blade an inch or two long. It branches and becomes more complex with age, forming first a lobed blade, and then as the blade wears away down to the coarser midrib, a plant with a branching stalk (the old midrib) and several blades as shown here. The blades look something like diminutive oak leaves because of the conspicuous veining. Each foliar expanse has a large central vein or midrib running the length of it. From this, numerous smaller, lateral veins arise pinnately and run a parallel route toward the margins of the frond. They may arise oppositely or alternately from the main vein. Each blade is about as thick as a piece of paper. The margins are irregularly toothed or lobed. Plants are deep red in the water, drying rosy pink to wine red. They stick to paper well and are another of the plants prized for making decorations on letters or cards.

Phycodrys and *Grinnellia* belong to a family of red seaweeds whose members are leaflike, with delicate veining. Veins in the leaves of higher plants are composed of greatly modified cells some of which conduct water and minerals to the leaf cells where green chloroplasts manufacture carbohydrate food through the process of photosynthesis, and others which return manufactured food to other parts of the plant. Seaweeds, in sharp contrast, need no conducting tissues. Water and the minerals it contains is all around them. In a seaweed like *Phycodrys*, all cells contain photosynthetic pigments and can make food independently. Veins of such algae conduct neither water nor food. They may be largely structural, but it is interesting that they simulate so closely the venation of a higher plant leaf.

Why does *Phycodrys* have such a beautiful rich red color, often found in leafy red algae growing in deep water? Light as it penetrates through seawater decreases in intensity and changes color. Red light diminshes increasingly with depth. Without red light, red objects look gray. Underwater photographs possess a blue-green cast in the absence of artificial lights. So why are deepwater seaweeds red? The red is due to a pigment (a phycobilin called phycoerythrin) which is very good at collecting blue-green light for photosynthesis. Remember the "color wheel" in elementary school—red is the complement of green and by extension something red is effective at absorbing the deeply-penetrating green light. Another strategy of deepwater seaweeds (and seaweeds living in shaded places such as *Plumaria*) is to increase the concentration of their photosynthetic pigments.

Grinnellia americana X 0.60

Grinnellia americana Ceramiales

Derivation: honors Henry Grinnell of New York.
Range: East Coast north to Massachusetts.

The first author has watched the underwater seasonal succession of seaweeds in the vicinity of Scraggy Neck, Buzzards Bay over the years. The sudden appearance and growth of *Grinnellia* in early August is a striking feature of the summer succession. It is a conspicuous plant, attached in shallow, warm water to the merest solid substrate such as a small shell or stone. It grows rapidly. The full-grown plant may reach almost two feet in length in particularly favorable locations. Even an average plant is spectacular. It consists of a single, tissue-paper-like, rosy pink blade, eight inches to a foot tall, and several inches broad. The blade expands widely immediately above a very short, thin stalk. The general outline of most blades is broadly oval, though the upper end may be somewhat pointed. Running more than two-thirds the length of the blade is a distinct midrib. Lateral veins as found in *Phycodrys* are lacking. The margin of the blade has more expanse than the central part, so it undulates in broad ruffles in the water and overlaps itself in deep folds when you try to spread a specimen out on a piece of paper. In general characteristics, the blade is like that of *Ulva*, but more delicate, veined, and of course red instead of green.

Plants of *Grinnellia* are of three kinds: male, and female (both haploid), and spore-bearing (the diploid tetrasporophyte). They can be distinguished by the naked eye at maturity, but it helps to have all three together for comparison. Sporebearing plants, a cluster of which are shown here, are dotted throughout with small splashes of deeper color. These spots are irregular in outline, and generally elongate. Female plants bear similar spots of color, but they are not as numerous, and mostly round in outline, as though the plant had been pricked all over with a pin point. Male blades are distinctly smaller and are not conspicuously marked.

Presence of all three types of plant in the same populations suggests that sexual maturity occurs rapidly, sexual reproduction takes place, and the resulting diploid, spore-bearing plants grow rapidly. If rapid development continues, and new sexual plants grow immediately from the first spores shed, a given population then represents several generations. Populations appear almost explosively in August. No plants can be found one day; the sea floor is heavily populated a few days later. Populations disappear as suddenly as they came, persisting only a month or six weeks. Where do they go? How do they live over winter? Which appears first in summer, the haploid or diploid phase? Do their reproductive cells come in from warmer waters to the south? Or do they overwinter here? Why don't they appear sooner in spring? Do they "wait" for the water to reach maximum summer temperature? Does their distribution correlate with particular water temperature? No one yet knows. The same questions could be asked for a number of other seaweeds.

Dasya baillouviana X 0.50

Dasya baillouviana

Ceramiales

Derivation: "hairy."
Range: Canadian Maritimes southward.

Professor Farlow, of Harvard, in 1881 called this, "a beautiful species, known to lady collectors by the name of chenille." Today the plant is no less beautiful than it was in Dr. Farlow's time. A well developed specimen may be two or three feet long, and in water its branches spread nearly a foot to the sides. It looks something like a great, open, peacock feather, ranging in color from intense red to wine or slightly purple, subject to some bleaching with age.

A typical plant is composed of a cylindrical central axis which bears numerous lateral branches, of approximately the same length. Most conspicuous is the "fur" on each of the branches. The "fur" consists of branched filaments that grow outward in great number all over the surface of the branch axis, and all to the same length. The "fur" is as densely colored as the axis. When you lift a plant from the water, the "fur" disappears because the delicate filaments collapse and lie against the surface of the branch. Some plants, as shown in the illustration, bear small dark "berries" among the filaments. These are reproductive structures. This plant is one that must be spread out in water in order to mount it effectively on paper. Like *Phycodrys* and *Grinnellia,* it adheres well to paper and keeps its color. These qualities lend it to use in making decorations and note cards. I have heard that enterprising students at Woods Hole have made pressed specimens of *Dasya* and *Grinnellia,* framed them with worm-ridden molding, and sold them at astronomical prices to tourists, who do not realize how readily available these plants are and how easily similar specimens can be made.

Dasya (only this species grows on our coast) lives in similar habitats as *Grinnellia* and, as discussed under that plant, appears and disappears from our waters seasonally in the same way. You will find large numbers of both plants in the wash at the edges of quiet, warm bays on the south side of the Cape during the month of August; rarely earlier, rarely later. Yet there are exceptions. *Grinnellia* has been reported to occur in quantity on the ice at Edgartown, and the Normans found a "second blooming" of *Dasya* far into the fall along the outer Cape.

Chondria baileyana on *Zostera* X 1.0

Chondria baileyana Ceramiales

Derivation: "cartilaginous."
Range: Virginia to Canadian Maritimes.

Despite its name, *Chondria* is not cartilaginous. Two species are common in Cape Cod waters and are often difficult to distinguish. *Chondria baileyana* is perhaps the most common. *Chondria capillaris* is the other one you are most apt to find. *Chondria* plants of either species are relatively easy to recognize, but it is difficult to say just what it is about them that makes this so. They are much-branching plants. The axes and branches are cylindrical and of relatively small diameter. A well-developed plant looks something like a coarse *Ceramium,* but without the banding. Never are they deeply colored. They are usually about the color of wheat straw, though young plants may have a purple hue. Perhaps their most distinctive feature is the nature of the ultimate branches. These are small, approximately equal in length, and more or less regularly distributed throughout the plant. The ultimate branchlets in *Chondria capillaris* are cigar-shaped or spindle-shaped, tapering distinctly toward both ends. Those in *C. baileyana* are most like baseball bats, blunt at the outer end. Frequently these terminal branchlets bear microscopic reproductive structures embedded in the surface, and this causes them to appear distinctly darker to the eye than the rest of the plant. When plants are dried on paper they become brown-yellow, and have a tendency to stain the paper the same color. On color alone, the average observer would place dried specimens of *Chondria* in the brown algae instead of among the reds.

Despite what is implied in its species name, *Chondria capillaris* is coarser than *C. baileyana.* The former is usually four to eight inches tall, bears many main axes spreading widely from the point of attachment, and is usually found on rocks and shells at and just below low tide line. *Chondria baileyana* rarely is more than a half foot tall, bears main axes and lateral branches somewhat in the form of a spruce tree, and grows mostly on eelgrass and larger algae in quiet warm waters.

Another species you may find is *Chondria dasyphylla.* It has thicker branches and much larger more club-shaped ultimate branchlets in which the tips are blunt. The whole plant is rigid enough that it does not collapse completely when taken out of water, as do the other two.

Polysiphonia elongata X 1.0

Polysiphonia elongata

Ceramiales

Derivation: "many-tubed."
Range: Long Island to Canadian Maritimes

Approximately a dozen species of *Polysiphonia* live on our Northeast Coast. The characteristics by which they are distinguished are mostly microscopic and not immediately obvious. In fact, the beginning observer may have considerable difficulty in distinguishing a *Polysiphonia* from a *Ceramium, Spyridia, Rhodomela,* or other nearly filamentous red algae. We will discuss three species on the next several pages.

A major characteristic of the genus, and one which is used at the same time to separate species, is the polysiphonous construction of the axes. Each axis is composed of a chain of axial cells surrounded by pericentral cells. To visualize this type of construction conceive a filament (single chain of cylindrical cells, ends butted together) laid lengthwise on this page. Conceive further that each cell of the filament divides in such a way that four cells come to lie lengthwise against it, all around it. The outside cells are narrow cylinders, each the same length as the central cell from which it divided. The end walls of the outside cells are butted against the end walls of those above and below it. They touch lengthwise with the ones on either side, and on the inside they touch lengthwise the central cell. This is shown well in the magnified drawing (as it would appear under the microscope) on the next page. Each axis of *Polysiphonia elongata* is constructed in this way. The original cell in the middle is called a central cell, and the ones around it, pericentrals. This type of construction is termed polysiphonous, and is found also among the brown algae in *Sphacelaria,* and with some modification, in *Spyridia* and *Rhodomela* among the reds.

Polysiphonia elongata grows attached to rocks and stones in relatively deep water. It is perennial, persisting from season to season, but its period of most active growth is in late winter and early spring. At this time it produces delicate filamentous outgrowths from its branches. These are later lost or worn away, and a plant collected in mid summer is apt to lack them altogether.

The main axis and branches of *Polysiphonia elongata* continue to increase in size and coarseness with age. This comes about through the process of cortication. Small, randomly oriented cells bud out from the pericentral cells in the older parts of the plant and form a layer of increasing thickness over the surface of the axis. As the layer of cortication thickens, it increasingly obscures the basic polysiphonous construction of the plant. This is why it is best to look among the younger parts of the major branches to check whether polysiphonous construction is present.

Polysiphonia nigrescens X 0.65

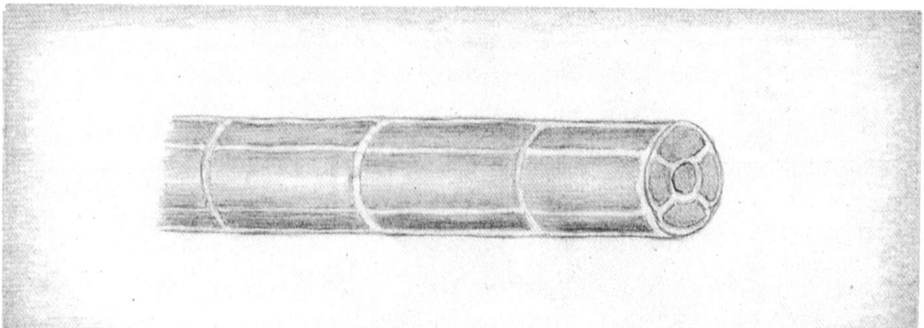

Polysiphonous construction greatly enlarged; 4 pericentral cells

Polysiphonia nigrescens

Ceramiales

Derivation: "many-tubed."
Range: North Carolina to Labrador.

The larger illustration on the opposite page shows *Polysiphonia nigrescens* with distinct main branches, clustering of the terminal branchlets, and the presence typically of sixteen pericentral cells (may range 8-20). This species may be quite dark in color (brownish red to almost black) as collected, and becomes back when dried.

Separating the species of *Polysiphonia* begins with determining how many pericentrals there are (see *Polysiphonia elongata*, preceding page). The largest group of species of *Polysiphonia*, divided on the basis of number of pericentral cells, contains those with four pericentral cells, diagramed in the smaller illustration here as it would appear in dissection under a microscope. Both *Polysiphonia elongata* and *P. urceolata* belong in this group.

Polysiphonia urceolata (not illustrated) is somewhat easier to recognize than other species of *Polysiphonia* by its lack of a distinct main axis, intense wine red color, branches that are soft and flexuous in texture, and a base with creeping branches spreading over the substrate. In contrast to *Polysiphonia elongata*, this species never forms cortication on the surface of its branches, which therefore remain more delicate in appearance throughout the life of the alga. It is a perennial seaweed, but most of the branching, erect part of the plant dies back and disappears in late summer, especially along the southern shores of Cape Cod and at the Islands. The matted holdfast part of the alga remains alive, however, and a new branching plant grows forth from it again during late winter and early spring of the following year.

The number of pericentral cells is constant for a given species of *Polysiphonia*, and throughout the individual plant. If your means of magnification is good enough to make the pericental cells visible, remember that you will see only about half of them from one side as you look down on an axis The common species of *Polysiphonia* in the Cape Cod area, and the total number of pericentrals for each are as follows: *P. harveyi*, 4; *P. urceolata*, 4; *P. elongata*, 4; *P. fibrillosa*, 4; *P. subtillissima*, 4; *P. nigra*, 12; *P. nigrescens*, 16; and *P. lanosa* 20-24. The fact of polysiphonous construction, if not the exact number of pericentral cells, can usually be determined with a good hand lens, but it is about at the level of resolution. Remember to look among the younger parts of the main branches. If you get an impression of transverse plus longitudinal markings, you probably have a *Polysiphonia*. *Ceramium* gives only transverse bands. *Spyridia* and *Rhodomela* have patrticular gross characteristics which will help distinguish them.

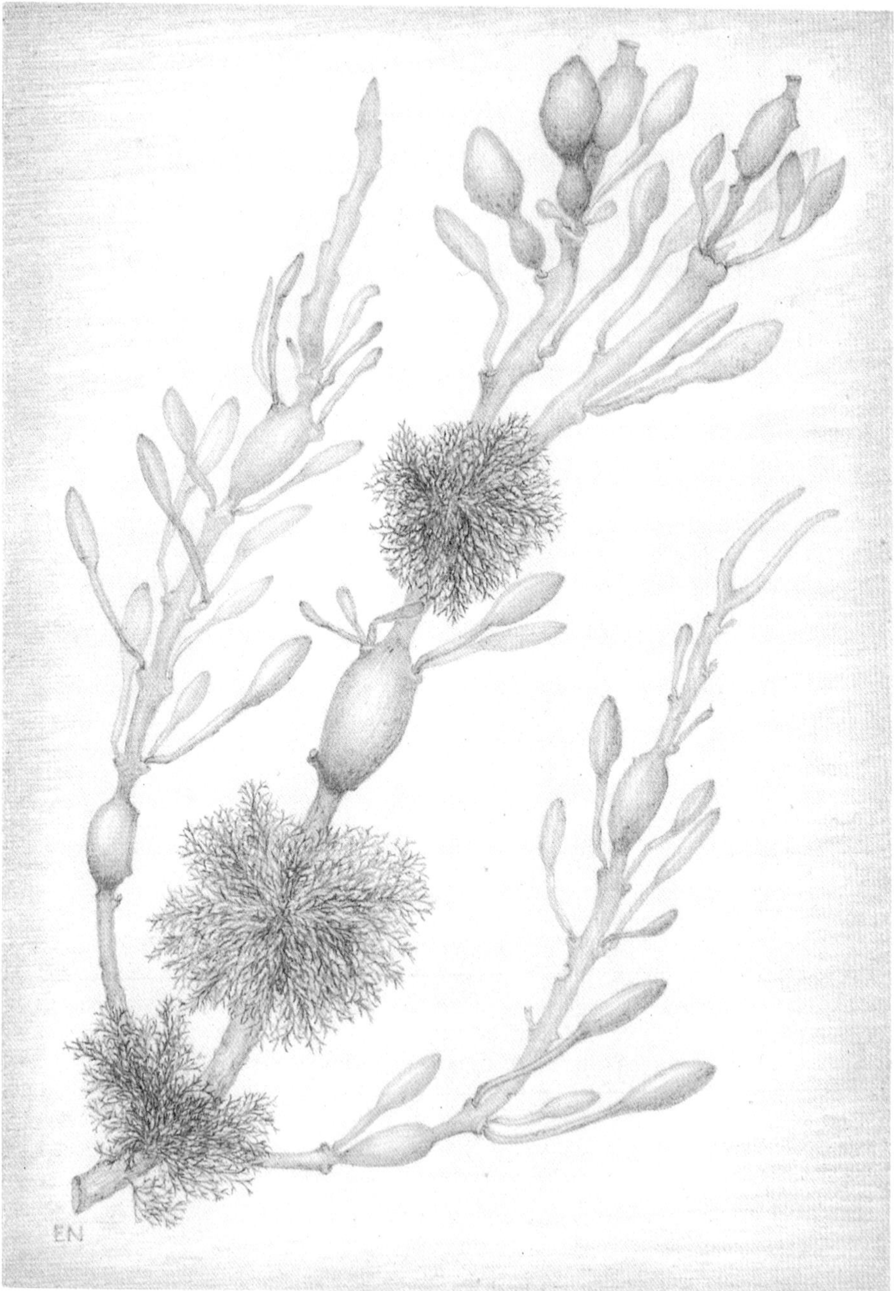

Polysiphonia lanosa on *Ascophyllum* X 0.60

Polysiphonia lanosa

Ceramiales

Derivation: "many-tubed."
Range: New Jersey northward.

Polysiphonia lanosa might be called the contrary *Polysiphonia*. It is coarser than most species of *Polysiphonia*, does not collapse on being removed from water, has more pericentral cells (20) than any other *Polysiphonia* in the Cape Cod area, is perennial, while many species are annual, and has a very specific habitat, while most species may be found on more than one substrate.

Polysiphonia lanosa grows only on *Ascophyllum nodosum*. Look for dark, almost black, bushy growths, one to three inches in extent attached to *Ascophyllum* and consisting of repeatedly and densely branched, stiff, wiry axes. The majority of branches on each plant end at about the same level so a whole, well-developed plant is nearly spherical and looks something like a clipped box plant. These characteristics are shown well in the illustration. Sometimes only a few individuals are scattered on the *Ascophyllum* "ropes," but often they occur in such quantity as nearly to obscure the brown alga.

The specificity of the host-epiphyte relationship between *Ascophyllum nodosum* and *Potysiphonia lanosa* is surprising. We have never found *Polysiphonia lanosa* on any other substrate. Why do the reproductive cells of this red alga, floating about in the ocean water, attach only to this particular brown alga? If *Ascophyllum* is easy to attach to, why isn't it loaded with other epiphytes? It may have a few, but it rarely supports a dense growth of anything but *Polysiphonia lanosa*. Part of the answer may well be that *Polysiphonia lanosa* is able to resist the effects of drying as the tide goes out. Many red algae cannot.

Specificity of host-epiphyte relationship varies from none to absolute. Many seaweeds attach to anything solid that time and tide makes available. Others make more specific requirements. Certain cyanobacteria grow in the relatively loose filamentous tissue of *Nemalion*, but not in firmer seaweeds. Others dissolve lime and can grow within the substance of mollusk shells. (To see this, look for old clam shells which are green or bluegreen, and from which the color won't scrape off). The sexual stage of *Ectocarpus siliculosus* apparently has a more specific epiphyte relationship than the other stage in its life cycle.

Epiphytism may lead in evolution to parasitism, in which the host provides not only substrate but also nourishment, and the parasitic plant loses its photosynthetic color. Some red algae are parasites. Among the red algae worldwide, about ninety percent of the parasites occur only on hosts that are very closely related to the parasites (same botanical family). Answers to the question of how these plants make themselves biochemically compatible with their closely related hosts might have significance for human medicine in suggesting how to improve the success with which organs may be transplanted from one individual to another.

Rhodomela confervoides X 1.0

Rhodomela confervoides

Ceramiales

Derivation: "red-black."
Range: New Jersey northward.

Rhodomela is not easily distinguished from *Polysiphonia*, because the principal difference has to do with the pericentral cells (see *Polysiphonia*) which are just visible with hand lens under the best of circumstances. *Rhodomela* possesses six or seven pericentral cells, but each pericentral cell becomes obscured in appearance by dividing further into smaller, irregular cells, the outlines of which are not obvious under a hand lens.

Rhodomela confervoides grows on stones below low water forming a plant a half foot to a foot and a half tall. Each plant consists of several coarse wiry axes about 1/16th inch thick, bearing delicate branches. It varies from dull red to almost black in color. In winter and early spring, the younger parts of these perennial plants are densely clothed with clusters of long, soft, fine branchlets. They are shown well in this illustration which depicts a plant in late winter condition. The clusters are shed as the season progresses, leaving the main axes more or less naked and giving the plant a vastly different appearance. During summer, plants collected in the wash are usually more or less denuded and puzzling as to identity.

Rhodomela confervoides also occurs in abundance on the British and northern European shores of the Atlantic, as far south as the Iberian Peninsula. In possessing this distribution, it is similar to a large number of other common seaweeds of the northern New England coast. In fact, the seaweed flora of the American shores north from Cape Cod bears greater similarity to the European coasts of the North Atlantic than it does to the seaweed floras of the other major sections of the American coast. Our Pacific shores have a seaweed flora more like that of Japan and Asia than that of Cape Cod. The warm water flora of localities from the south shore of Cape Cod southward grades into a tropical flora characteristic of the Caribbean and the Gulf Coast.

Thus it is that an interested person can study three greatly differing floras of seaweeds (Atlantic, Pacific, and subtropical) without leaving the shores of the continental United States, finding a variety which occurs in few other parts of the world. Cape Cod, lying at a boundary between northern and southern elements, is particularly well situated for the purposes of those who wish to study seaweeds.

OUTLINE BY SPECIFIC CHARACTERISTICS

This outline is presented as an aid to identifying seaweeds. Here the seaweeds discussed in this book are separated into small groups on the basis of specific characteristics. By this means the number of descriptions and illustrations with which a given specimen must be compared can be reduced to just those with the characteristic in question. For reasons explained elsewhere, a list of this sort cannot be taken as absolute, and may not work satisfactorily in every case, but it should be helpful if used with an appropriate amount of judgment and discretion.

GREEN ALGAE

Plants small, felt-like en masse — *Prasiola*, 38

Plants filamentous or apparently filamentous

- Filaments unbranched (hand lens) — *Chaetomorpha*, 40, 42; *Ulothrix*, 24
- Filaments branched* — *Bryopsis*, 48; *Cladophora*, 44; *Spongomorpha*, 46

Plants tubular (may be flattened in part)

- Plants hollow — *Enteromorpha*, 28-34
- Plants solid — *Codium*, 50

Plants sheet-like — *Enteromorpha linza*, 34; *Monostroma*, 26; *Ulva*, 36

*Some very delicate species of *Enteromorpha* may come out here.

BROWN ALGAE

Plants filamentous or apparently so — *Ectocarpus*, 52,55; *Elachista*, 64; *Isthmoplea*, 80; *Pilayella*, 56; *Sphacelaria*, 58

Plants encrusting

- On rocks, wood, etc. — *Ralfsia*, 90
- On other seaweeds — *Myrionema*, 62

Plants lumpy, like popcorn — *Leathesia*, 66

- Plants tubular or cylindrical or cordlike throughout; unbranched
  - Plants mostly solid — *Chorda*, 96, 98
  - Plants hollow, somewhat inflated — *Asperococcus*, 86; *Scytosiphon*, 94
- Plants cylindrical throughout; branched; lacking air bladders
  - Axes soft, wormlike — *Eudesme*, 68
  - Plant bushy, erect or lax but not wormlike
    - Ultimate branchlets whorled, brush-like — *Cladostephus*, 60
    - Branches not sharply distinct from axis bearing them — *Acrothrix*, 74; *Chordaria*, 72; *Desmarestia*, 76, 78; *Dictyosiphon*, 88; *Sphaerotrichia*, 70
- Plants cylindrical to flattened; branched; air bladders present — *Ascophyllum*, 120
- Plants cylinclrical to flattened; branched; air bladders lacking — *Desmarestia*, 76, 78
- Plants with clearly differentiated blades (flat foliose expanse)
  - Blades with midrib
    - Holdfast unbranched; axes repeatedly branched — *Fucus*, 110-118; *Sargassum*, 124
    - Holdfast branched; axes unbranched — *Agarum*, 106; *Alaria*, 108
  - Blades lacking midrib
    - Blades with regular holes — *Agarum*, 106
    - Blades with splits; segments ribbon-like — *Laminaria*, 100-104
    - Blades single, simple
      - Holdfast unbranched; no clearly differentiated stalk — *Petalonia*, 92; *Punctaria*, 82, 84
      - Holdfast branched; stalk clearly differentiated from blade — *Laminaria digitata*, 104

RED ALGAE

- Plants encrusting — *Hildenbrandia*, 142; *Phymatolithon*, 144
- Plants with lime secreted in tissues — *Corallina* 146, *Phymatolithon*, 144
- Plants with distinct hook-like tendrils or "pigtails" in branches (may not be numerous) — *Bonnemaisonia*, 134; *Cystoclonium*, 158; *Hypnea*, 162

Plants delicate, flat, with branches like the teeth of a double sided comb
Plumaria, 188

Plants filamentous throughout or apparently so; soft, lax, collapsing out of water; possessing none of the special characteristics above

- Cells of filaments visible to unaided eye — *Griffithsia*, 186
- Cells corticated; axes banded — *Ceramium*, 192
- Cells smaller — *Antithamnion*, 183; *Audouinella*, 130; *Bangia*, 126; *Callithamnion*, 181; *Polysiphonia*, 202-206; *Seirospora*, 184; *Spermothamnion*, 182; *Trailliella*, 135

Plants filamentous in outer portions, or apparently so, collapsing out of water; main axes thicker, cylindrical, not collapsing — *Dasya*, 198; *Polysiphonia*, 202-206; *Rhodomela*, 208; *Spyridia*, 190

Plants delicate but wiry throughout; axes less than 1/16th inch diameter, cylindrical — *Ahnfeltia*, 172; *Gelidium*, 138; *Polyides*, 156; *Polysiphonia lanosa*, 206

Plants larger, coarse; axes cylindrical, branching, soft, cartilaginous, or somewhat stiff, but not wiry

- Plants soft and wormlike — *Nemalion*, 132; *Platoma*, 150
- Plants fuzzy; ultimate branchlets filamentous and of equal length — *Dasya*, 198
- Axes clearly segmented to unaided eye — *Ceramium*, 192; *Champia*, 178

Plants bush-like in general shape; a major axis may or may not be obvious; with none of the special characteristics above — *Agardhiella* 160, *Chondria*, 200; *Cystoclonium*, 158; *Dumontia*, 148; *Gloiosiphonia*, 152; *Gracilaria*, 174; *Lomentaria*, 176; *Polyides*, 156

Plants blade-like in whole or in part	
Plants flat throughout; no distinct stalk	
Blades greatly dissected	*Callophyllis,* 154
Blades clearly veined	*Grinnellia,* 196 *Phycodrys,* 194
Blades not greatly dissected, nor veined	
Plants sparsely branched	*Porphyra,* 128 *Palmaria,* 140
Plants moderately branched	*Chondrus,* 168, 171 *Mastocarpus,* 166 *Gracilaria,* 174
Blades flat; plants clearly stalked	
Blades veined	*Phycodrys* 194,
Blades not veined	*Chondrus,* 168, 171 *Dumontia,* 148 *Mastocarpus,* 166 *Gracilaria,* 174 *Phyllophora,* 164

GLOSSARY

Although specialized terminology available for application to the algae has been avoided in this book whenever an equivalent term in ordinary English would suffice, a few botanical or phycological terms are unavoidable. They are always defined in context at some location in the text, sometimes at several places. Nevertheless, the nature of the book suggests that it will be entered directly at almost any page and it has not been possible to repeat a definition at every point of use of a special term. The glossary below defines a few terms and directs the reader to a definition in textual context of the others.

BIBLIOGRAPHY

Berrill, Michael and Deborah Berrill. *The North Atlantic Coast.* Sierra Club Books, San Francisco. 1981.

Bird, Carolyn J. and Jack L. McLachlan. *Seaweed Flora of the Maritimes, 1. Rhodophyta - The Red Algae*. Biopress, Bristol, England. 1992.

Conkling, Philip W. (ed.) *From Cape Cod to the Bay of Fundy.* The MIT Press, Cambridge.

Schneider, Craig W. and Richard B. Searles. *Seaweeds of the Southeastern United States.* Duke University Press, Durham, North Carolina. 1991.

Sze, Philip. *A Biology of the Algae.* WCB/McGraw Hill Publishers, Dubuque, Iowa. 3rd ed. 1997.

Taylor, William Randolph. *Marine Algae of the Northeastern Coast of North America*. University of Michigan Press, Ann Arbor. 2nd ed. 1957.

Villalard-Bohnsack, Martine. *Illustrated Key to the Seaweeds of New England.* The Rhode Island Natural History Survey, Kingston. 1995.

NOTE

Phycologists do not always agree on the appropriate name for some algae. Alternative names found in some of these references are:

Monostroma oxyspermum = *Ulvaria oxysperma*
Spongomorpha arcta = *Acrosiphonia arcta*
Scytosiphon lomentaria = *S. simplicissimus*
Agarum cribosum = *A. clathratum*
Phyllophora truncata = *Coccotylus truncatus*
Ceramium rubrum = *C. nodulosum*
Polysiphonia nigrescens = *P. fucoides*
Polysiphonia urceolata = *P. stricta*

INDEX

This index is devoted almost exclusively to seaweeds (or other organisms) by name. Italics are used for organism names. Refer to the Glossary (page 215) for additional terms. Boldface indicates a main entry. Main entries for species are almost always accompanied by an illustration on the preceding page. Pagination for Classes is given in the Contents, p. 7.